Soil, Fertilizer, and Plant Silicon Research in Japan

Soil, Fertilizer, and Plant Silicon Research in Japan

Jian Feng Ma
Kagawa University, Kagawa, Japan

Eiichi Takahashi
Kyoto University, Kyoto, Japan

2002
ELSEVIER
Amsterdam-Boston-London-NewYork-Oxford-Paris
San Diego-San Francisco-Singapore-Sydney-Tokyo

ELSEVIER SCIENCE B.V.
Sara Burgerhartstraat 25
P.O. Box 211, 1000 AE Amsterdam, The Netherlands

First edition 2002

Library of Congress Cataloging in Publication Data
A catalog record from the Library of Congress has been applied for.

British Library Cataloguing in Publication Data
A catalogue record from the British Library has been applied for.

ISBN: 0 444 51166 0

Transferred to digital printing 2007

Preface

Silicon (Si) is the second most abundant element in the earth's crust and all plants rooting in soil contain significant amounts of Si. However, due to its universal existence and lack of obvious deficiency symptoms, the impact of Si on the growth of plants has not been paid much attention for a long time. This case resembles gravity; nobody recognizes its significance in daily life although we can not stand on the ground without it.

The significance of Si on rice production was recognized in Japan about 80 years ago. Rice is the most important crop in Japan and intensive scientific agricultural researches on rice have been carried out since the Meiji Revolution. In a time without pesticides, the main threats for rice production were diseases and pests. Research findings on the effects of Si in controlling diseases and pests attributed to the growth of interest among agronomists. Furthermore, Si was found to be important in improving degraded paddy fields and autumn decline of rice. Numerous trials conducted nationwide on Si led to a conclusion that a high content of Si is required for the healthy growth of rice, a typical Si-accumulator, and for stable rice productivity. Nowadays, Si is considered as an agronomically essential element in Japan.

On the other hand, after World War II, with the rapid development of the steel and iron industry, a large amount of slag, a by-product, resulted. Slag mainly contains calcium silicate and various field trials showed that slag can be utilized as a Si source for rice. In 1955, slag, which was applied in Europe as a liming material, was recognized as a Si fertilizer in Japan for the first time in the world.

Under such a background, Japanese scientists have been conducting intensive studies on Si from the viewpoints of soil, fertilizer, and plant nutrition and have made great progress. Unfortunately, most reports on Si are in Japanese, which makes access to these achievements by non-Japanese difficult. With the increasing global interest on Si, there are eager requests from overseas colleagues to introduce the studies on Si conducted in Japan. In this book, we have tried to introduce the major achievements on Si research concerning soils, fertilizers and plants in Japan. However, as there are numerous works on Si, not all of them could be included in this publication. Readers who are interested in a particular aspect should contact us without hesitation. We may be able to provide more information about Si research in Japan personally.

We hope that this book will be helpful for those interested in Si researches and will stimulate further researches on Si worldwide.

Jian Feng Ma

Faculty of Agriculture
Kagawa University
Mikicho, Kagawa
Japan

Eiichi Takahashi

Kyoto University
Sakyo-ku, Kyoto
Japan

Table of Contents

Chapter 1

Brief history of silicon research in Japan—Birth of silicate fertilizer

Silicon (Si) is the second most abundant element, both in terms of weight and number of atoms, in the earth's crust. Because of its strong affinity with oxygen, Si in nature always exists as silica (Si dioxide) or silicates, which are combined with various metals. Silicon dioxide comprises about 60% of the earth's crust, and it occupies more than 50% of the soil and the concentration of Si in soil solution in the form of silicic acid is between 3.5 and 40 mg Si L^{-1} (Marschner 1995). Therefore, all plants rooting in soil contain Si in their tissues. However, because of its universal existence, researchers did not pay much attention to the impact of Si on plant growth. In his article on Si nutrition, Sachs (1862) first asked "whether silicic acid is an indispensable substance for those plants that contain silica, whether it takes part in nutritional processes, and what is the relationship that exists between silicic acid and the life of the plant." However, Sachs concluded that Si was insignificant for the nutritional process of maize when he found that water-cultured maize which contained 0.3% Si did not show any abnormality in growth compared with maize containing 9% Si. Since then many studies have been carried out on the effect of Si on plant growth.

In Japan, Si research started at the beginning of the twentieth century. In 1917, Onodera reported that the Si content of rice leaves infected with blast disease was lower than that of healthy leaves. This is the first report on Si research published in a scientific journal of agronomy. Later, several papers on the relationship between Si and blast disease were published. For example, it was found that blast disease-resistant cultivars contain a larger amount of Si than blast-sensitive cultivars (Miyake et al., 1922) and application of silicate increased the resistance of rice to blast disease (Kawashima, 1927; Miyake et al., 1932).

Researches on the physiological role of Si in rice were initiated by Ohkawa (1936-1942) and Ishibashi (1936-1939). They found that Si deficiency inhibits the growth of rice, especially the increase in empty grains significantly reduced the yield. They also found that the effect of Si application was more obvious under heavy application of nitrogen fertilizers and that the damage due to brown spot and blast was alleviated by Si application.

Thus it became clear in the late 1930s that Si is important for healthy growth and high grain yield of rice, but the results of these laboratory researches were not applied to the field. Because the Si abundantly present in the soil is gradually solubilized with weathering and Si could also be supplied from irrigation water and compost, it was considered that addition of Si to soil was unnecessary. In addition, appropriate Si fertilizers were not available at that time.

However, from the field trials of akiochi (autumn decline rice growth which happened in degraded paddy soils) in the 1940s, it was recognized that Si in the soil was not sufficient for the healthy growth of rice (Mitsui et al., 1948; Hashimoto et al., 1948).

Rice is the most important crop in Japan where it has been cultivated successively on submerged soil for some 1700 years. Rice cultivation is characterized by continuous cropping without any injury. This cultivation system is different from that of Europe, where the main crops are upland winter cereals. The continuous cropping system suits Japan where only a little arable land is available per person. Paddy field, a special cultivation environment makes continuous cropping of rice possible (upland rice can not be cultivated continuously).

Compared with upland field, paddy field gets lots of irrigation water. The total amount of irrigation water during the rice growth period reaches 14 thousands tons per hectare. Some of the water evaporates from the leaves and surface of paddy fields, and the remaining water is percolated through the soil leaching its soluble components. During the waterlogging period, paddy soil is always washed with water, resulting in leaching of both toxic compounds and nutrient bases such as Ca, K and Mg.

In addition, because the soil surface is covered with water and subsequently shut off from air, the oxygen in soil is consumed and the soil condition becomes strongly reductive. The iron in soil is then reduced and subsequently solubilized, resulting in eluviation of Fe to subsoil. Similarly Mn and Si are also eluviated. Sulfate ion in soil is also reduced to H_2S, which is converted to non-toxic FeS with Fe, but when there is not enough active Fe the rice roots will be damaged by H_2S.

In degraded paddy soils nutrients such as the bases, Fe, Mn, and Si are leached out and rice roots can be seriously damaged by H_2S. In these soils, "akiochi" of rice easily happens. "Akiochi" is a phenomenon in which the growth of rice is good until the end of summer, but thereafter the lower leaves become wilty and brown spots occur, finally resulting in low yield. Such paddy soil can also be used for continuous cropping, but the productivity decreases year after year.

At the time of food shortage after the Second World War, it was an important project to improve degraded paddy soils and therefore many field trials were conducted. The results indicated that supplement of Si in addition to bases and Fe is also important for the improvement of the productivity in declined paddy soils. On the other hand, it was found that slag (calcium silicate as a major component), a by-product from the iron industry that was developing rapidly, could be used as a potential silicate fertilizer. In 1952, the Ministry of Agriculture, Forestry and Fisheries of Japan therefore started field trials on slag at nationwide agricultural experiment stations.

The results of these trials revealed the beneficial effects of slag at various places and in 1955 the Ministry of Agriculture, Forestry and Fisheries decided to put Si on the fertilizer list. Slag that contains a sufficient amount of Si available to the plant, but free of toxic components was recognized as "silicate fertilizers" and the standard of silicate fertilizer was decided. Slag is also utilized in European and American countries, but it was used as lime materials. Japan is the first country to use slag as a Si source for crops.

Silicate fertilizer was born in Japan for the following reasons

1. Rice, the most important crop in Japan, is characterized by high accumulation of Si in the plant;
2. In Japan, rice is usually cultivated at a high density with heavy application of nitrogen fertilizer;
3. Si-deficient soils such as degraded paddy soils are widely distributed;
4. Cheaper silicate fertilizers like slag are provided from iron industry; and
5. Return of the main Si source rice straw to the paddy soil is a gradually decreasing practice mainly because of labor shortage.

The birth of silicate fertilizers is closely linked with this specific nature of Japanese agriculture.

Research on Si has been actively conducted in various areas such as soil, fertilizers and plant science, as will be detailed in each chapter.

Chapter 2

Silicon sources for agriculture

2.1. SILICON SUPPLY FOR PADDY RICE FROM NATURAL SOURCES

Natural sources of Si for rice are irrigation water and soil. The amount of Si supplied varies with the parent material and geology of river basin.

2.1.1. Irrigation water

Paddy soil is irrigated with an average of 14,000 tons of water per hectare during the growth period of rice. Therefore, Si in irrigation water has an important impact on rice production. Kobayashi (1954) collected 116 samples of rice straw from various places in Japan and found a positive relationship between Si concentration in rice straw and in irrigation water (Figure 2.1).

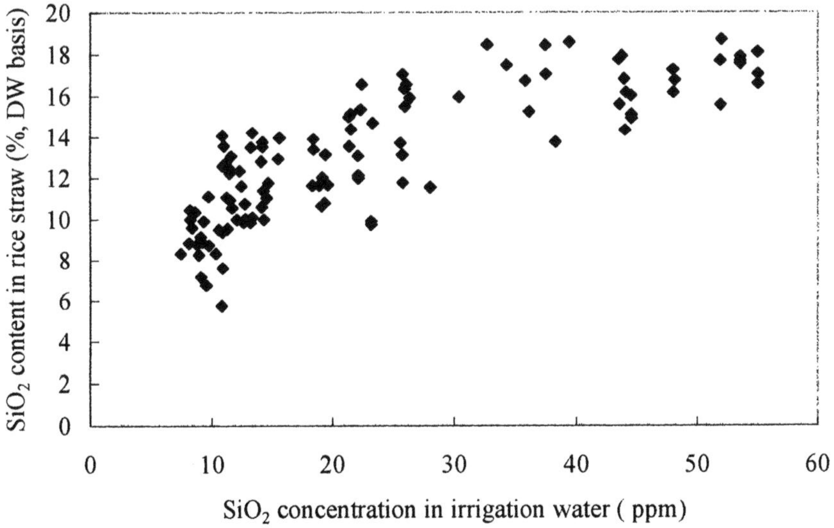

Figure 2.1. The relationship between Si concentration in irrigation water and Si content in rice straw.

Regions		No. of rivers	SiO_2 (ppm) Maximum	Minimum	Average
I	Hokkaido	40	49.7	10.2	27.0
II	16 prefectures	166	61.5	8.0	21.9
III	6 prefectures	34	27.7	7.6	14.4
IV	16 prefectures	76	31.7	4.1	13.6
V	8 prefectures	64	54.6	10.9	30.9
		380	61.5	4.1	21.6

Figure 2.2. Average SiO_2 concentration of river waters in five regions of Japan.

Table 2.1
The effect of geology of basin on SiO_2 concentration in river water

Geology of basin	Number of rivers	SiO_2 concentrations (ppm)
Paleozic (aqueous rock)	5	12.5
Mesozoic (aqueous rock)	4	11.0
Granite	5	13.9
Volcanic ash	7	47.5

He also collected river water used for irrigation and measured its Si concentration by a colorimetric method (Kobayashi, 1961). As shown in Figure 2.2, among 380 rivers investigated, the lowest concentration of SiO_2 was 4.1ppm, and the highest was 61.5 ppm with an average of 21.6 ppm (for details, refer to Appendix I). If rice is irrigated with 14,000 tons water per hectare, it is calculated that an average of 300 kg SiO_2 per hectare is supplied to rice from irrigation water annually.

In Figure 2.2, the rivers in I, II and V regions have a high Si concentration. This is related to existence of volcanoes in these areas. The Si concentration of river water also varies with geology of basin. River water originating from aqueous rock and granite usually has a low Si concentration, while that from volcanic ash has a high Si concentration (Table 2.1).

2.1.2. Soils

The Si supply capacity of soils varies greatly with paddy field. In 1955, the Ministry of Agriculture, Forestry and Fisheries of Japan made a nationwide survey on Si content in the flag leaf, using the data from 37,949 samples collected from various paddy fields. The Si content in the flag leaf varied widely with the region as shown in Table 2.2 (for details, refer to Appendix II). In about 5% of samples examined, the Si content was less than 7.5%, while 9% of samples showed the Si content higher than 23%. The Si content in the flag leaf reflects the Si supply power of paddy soils. In regions II and V, the Si content is high, while the Si content in regions III and IV is low. This trend is consistent with that for Si concentration in river waters reported by Kobayashi (Figure 2.2).

In the regions where the Si content in the leaf flag is lower than 7.5% and where the Si content is between 7.6-12.5%, there is a possibility of Si-deficiency in the soil. The application of Si fertilizers would be effective in these regions, mainly in regions III and IV (Table 2.2).

Table 2.2
Average SiO_2 content in flag leaf in the five regions of Japan in 1955

Regions	No. of sampling sites	Average SiO_2 content (%)	Percentage of samples with a SiO_2 content (%) of	
			<7.5 %	7.6−12.5 %
I	278	13.6	2.9	39.2
II	15,902	17.8	1.2	12.3
III	8,059	13.0	9.5	40.3
IV	9,102	13.2	9.6	36.3
V	4,608	17.5	1.2	13.0
Whole land	37,949	15.6	5.4	24.2

Each sample (flag leaves) was collected at different sites in each region.

Imaizumi and Yoshida (1958) summarized the Si-supplying capacity of soils with different parent materials, based on the field trial data from nationwide agricultural experiment stations. As shown in Table 2.3, volcanic ash (Tochigi) and shale (Gifu)-originating soils have a high Si-supplying capacity. Furthermore, it was found that new volcanic ash soil is rich in water-soluble Si, but the Si-supplying capacity of volcanic ash soils decreases with aging because of desilicating process. Lower Si concentration in volcanic ash soils of Gifu and

Table 2.3
The relationship between parent material and SiO_2 supplying capacity of soil

			SiO_2 content			
			Range		Average	
Parent material	Locality	Sample size	Rice straw (%, DW)	Soil available SiO_2[a] (mg/100g soil)	Rice straw (%, DW)	Soil available SiO_2[a] (mg/100g soil)
Volcanic ash	Tochigi	3	15.6-18.1	22.7-31.0	16.8	28.2
Shale	Gifu	8	14.3-19.6	13.3-27.7	16.3	20.0
Quartz porphyry and granite	Gifu	9	5.2-13.1	5.0-20.7	9.1	9.5
Volcanic ash (desilicated)	Gifu	5	9.6-12.3	5.8-12.2	10.5	9.2
Volcanic ash (desilicated)	Yamagata	8	4.9-10.3	5.9-11.6	7.5	8.3
Granite	Nagano	6	4.9-9.8	3.6-7.6	6.7	5.7
Granite	Yamagata	8	4.7-8.3	2.5-10.4	7.2	5.3
Peat	Yamagata	2	4.4-5.8	--	5.1	4.9

[a]Acetate buffer (pH 4.0) soluble SiO_2.

Table 2.4
Balance of Si measured with a lysimeter charged with various kinds of paddy soils (SiO_2 g m^{-2}, Hokuriku Agricultural Experiment Station, 1968)

Soil texture	Supply from irrigation water (A)	Loss in Percolating water (B)	Uptake by Rice plant (C)	From irrigation water (A-B)	From Soil (C-(A-B))
Sandy soil	52.14	45.72	63.75	6.42	57.33
Heavy clay soil	50.42	29.81	139.75	20.61	118.64
Peat soil	54.49	34.34	93.50	20.15	73.35

Yamagata prefectures than that of Tochigi is attributed to different age of volcanic ash. Soils derived from quartz porphyry and granite, and peat have low capacity of Si supply.

A balance sheet for Si taken up by rice and for Si supplied from irrigation water was calculated. The average amount of Si uptake by rice was estimated to be 950 kg SiO_2 per hectare, of which 260 kg SiO_2 originated from irrigation water. This fact suggests that 27% Si in rice is supplied from irrigation water, and the remaining Si is from soil. Therefore, Si in soils is a major natural Si source for rice.

The results of a lysimeter experiment conducted in Hokuriku Agricultural Experiment Station (1968), showed that except in sandy soil, 40% of Si in irrigation water is taken up by rice, which contributed to 10 to 20% of total Si of rice (Table 2.4).

2.2. SILICON SUPPLY FROM ORGANIC AND INORGANC FERTILIZERS

2.2.1. Compost

2.2.1.1. Application rate in the past and the present

Before the introduction of silicate fertilizers, the main source of Si supply was compost. For example, compost production was about 20 million tons in 1927. Because of fertilizer shortage during World War II, the use of self-supplied fertilizer was encouraged, and compost production reached 60 million tons around 1945. Assuming that compost contained 5% SiO_2, 1 million (1927) and 3 million (1945) tons of Si annually for 3 million hectares of paddy soil, would be equal to 330 and 1000 kg SiO_2 per hectare, respectively.

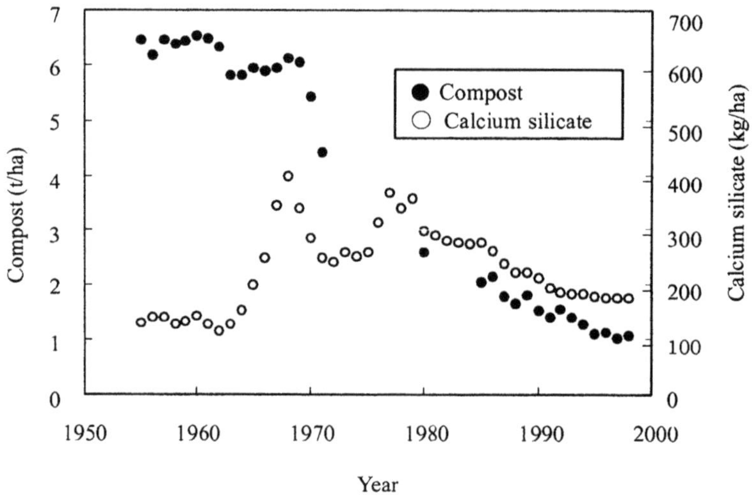

Figure 2.3. Changes in the amounts of Si supplied to the paddy field from compost and silicate fertilizer.

In 1955, when slag was introduced as silicate fertilizers, compost was applied at an average rate of 6.5 tons per hectare. However, from 1960, the rate of compost application gradually decreased, resulted in 4.5 tons in 1970, and less than 2 tons in 1990. Silicon supply from compost therefore decreased from 325 kg SiO_2 (1955) to 225 kg (1970), and to 100 kg (1990) per hectare. The decrease of Si supply from compost was supplemented by silicate fertilizers (Figure 2.3). The average rate of silicate fertilizer application per hectare increased from 150 kg in 1960 to 200 kg in 1965, and to more than 400 kg in 1968. Assuming that the Si content of silicate fertilizer is 25% SiO_2, 38 kg, 50 kg, and 100 kg SiO_2 would have been supplemented in 1960, 1965 and 1968 respectively. However, the rate of silicate fertilizer application has been gradually decreasing thereafter (Figure 2.3).

2.2.1.2. Short-term availability of Si in rice straw for rice plants

Rice straw used for preparation of compost contains up to 20% SiO_2. Ma and Takahashi (1989b, 1991a) reported the short-term availability of Si contained in rice straw for rice. They examined the availability as the solubility of Si in the rice straw. Rice straw containing 15.5% SiO_2 was successively extracted with distilled water at 40°c. This extraction method is the same as that for measurement of soluble Si in soil proposed by Takahashi as described in the next chapter. The concentration of Si dissolved in water was determined every week. The first and second extractions gave a concentration of about 100 ppm

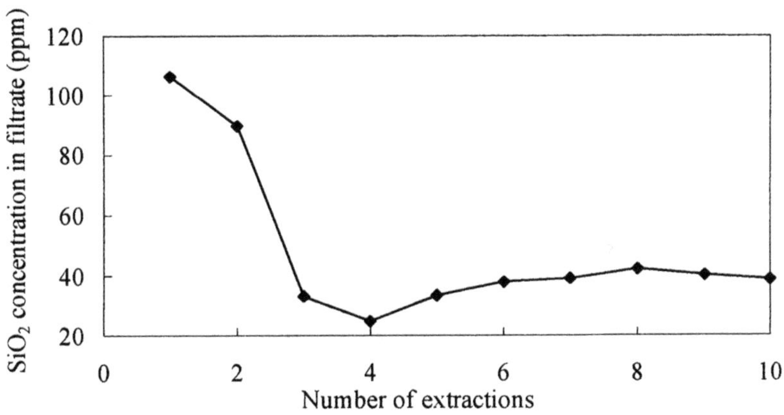

Figure 2.4. Solubility of Si in rice straw.

SiO_2 in the water (Figure 2.4). However, the solubility of Si in the later extractions decreased to about 40 ppm SiO_2 and remained at a similar level thereafter. The high concentration of Si during the first two extractions might have resulted from the presence of monomeric Si and Si with low polymerization in rice straw. They accounted for 4.1 and 3.6% of total Si for the first and second extraction, respectively. Most of the Si in rice straw showed solubility between that in silica gel and opal.

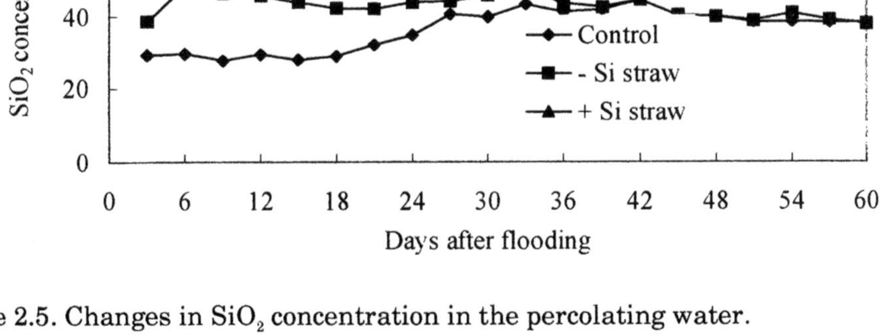

Figure 2.5. Changes in SiO_2 concentration in the percolating water.

Table 2.5
Total amount of silicon released during the experiment (mg SiO_2)

Treatment	Control	-Si straw	+Si straw
	380 (100)	448 (118)	830 (218)

Note: Figures in parentheses indicate the relative percentage of control.

The effect of rice straw on the solubility of soil Si was investigated under rice planting and non-planting conditions. As organic matter may affect the solubility of soil Si, rice straw without Si (-Si straw), prepared by cultivating rice hydroponically in the absence of Si, was used as a control. When rice plants were not planted, the Si concentration in the water percolating through the soil with +Si straw was 1.5-2 fold higher than that in the treatment with –Si straw (Figure 2.5), suggesting that Si in the straw is released (Ma and Takahashi, 1989b). The amount of Si released from the soil with +Si straw during the experiment (60 days) was 450 mg larger than that released from the soil without rice straw (Table 2.5), but this accounts for about 10% of Si contained in the +Si straw (4650 mg SiO_2). Addition of -Si straw also caused the Si concentration in the percolating water to increase (Figure 2.5). This increase is attributed to the increased solubility of Si in soil, resulting from the enhanced soil reduction.

The effect of rice straw on the amount of Si extracted from the soil determined at the end of the experiment, varied with the method of extraction. When acetate buffer was used to extract Si, no significant increase in the amount of extractable Si was found by the application of rice straw (Table 2.6). However, when Si was extracted by incubation under a submerged condition, about 40% increase in the amount of extractable Si by the application of +Si rice srtaw.

In the field under rice cultivation, the Si concentration in the percolating water decreased sharply compared with that in the percolating water without rice cultivation (Figures 2.5, 2.6), suggesting that rice roots absorb Si actively. Addition of +Si rice straw increased the Si concentration in the percolating water during the first two weeks but not thereafter (Figure 2.6).

Table 2.6
Amount of Si solubilized from soil with or without rice straw (mg SiO_2/100 g soil).

Treatment	Method of extraction	
	Acetate buffer	Incubation under submerged condition
Without straw	31.9	14.7
-Si straw	31.0	15.1
+Si straw	33.4	20.0

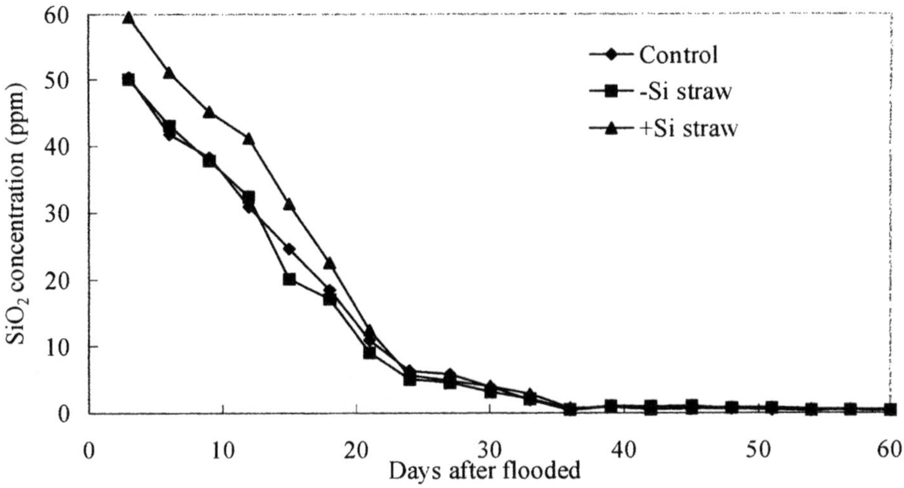

Figure 2.6. SiO_2 concentration in the percolating water during growth under different conditions (first cultivation).

Application of –Si straw did not affect the concentration of Si in the percolating water. Application of +Si straw increased Si uptake by rice (Table 2.7), while application of –Si straw did not affect Si uptake. The increased amount of Si uptake due to +Si straw application was 367.8 mg SiO_2 per pot, which accounted for 8% of Si contained in the +Si straw. Thus the availability of Si for the rice plant was low in the rice straw in the short-term experiments.

2.2.1.3. Long-term availability of Si in compost for rice plants

One example of the long-term effect of Si in compost is shown in Table 2.8. Compost was applied to the field from 1933 to 1973 (40 years) at Shiga Prefectural Agricultural Experiment Station. The summer crop was rice and the winter crop was wheat. The trial was done using 16 plots: 8 plots without compost (chemical fertilizer only) treatments, and 8 with compost (compost was

Table 2.7
Uptake of Si by rice from soil with or without rice straw

	Si uptake (SiO_2 mg/pot)		
	First cultivation	Second cultivation	Total
Without straw	826.2	441.2	1267.4
-Si straw	848.7	429.7	1278.4
+Si straw	1033.8	601.4	1635.2

Table 2.8
Contents (%) of Si and other elements in the straw of rice (first crop) and wheat (second crop) cultivated in the soils composted for 40 years and in non-composted soils (Report of Long-term Field Experiment (1933-1973) conducted at Shiga Prefectural Agricultural Experiment Station)

	Rice straw			Wheat straw		
	Composted Series, A	Non-composted Series, B	A/Bx100	Composted Series, A	Non-composted Series, B	A/Bx100
SiO_2	13.71	11.33	121	9.11	6.43	142
N	0.41	0.42	98	0.48	0.48	100
P	0.053	0.064	83	0.120	0.098	122
K	1.56	1.53	102	1.17	1.22	96
Ca	0.49	0.55	89	0.27	0.28	96

applied at a rate of 11.5 tons per hectare for summer crop, and 9.9 tons for the winter crop together with chemical fertilizers). In 1974, the year after the last compost application, rice and wheat were cultivated similarly, and the accumulative effect of compost applied for 40 years was investigated by monitoring yield, mineral content, physical and chemical properties of soil (Nakada, 1980).

Among the minerals investigated, the Si content in rice and wheat that had been grown in composted soil was higher than that in non-composted soil (Table 2.8). Table 2.9 shows that the Si content in both rice and wheat was higher in the treatments without N and P fertilizer application.

The amount of nutrients taken up by rice during the 40 years was estimated from the mineral content and dry matter of both straw and grain (Table 2.10). The amount of Si uptake by rice was much higher than that of the other minerals. It was 5-7, 16-23, 3-4, and 9-12 times as large as that of N, P, K and Ca, respectively. Furthermore, the uptake of nutrients, especially Si, was increased by compost application. Compared with the non-composted plots, 16 tons more SiO_2 per hectare was taken during the 40 years in composted plots. As 460 tons compost (estimated as 23 tons SiO_2) was applied during the 40 years, the Si taken up by rice accounts for 70% of the compost applied. If it was derived from the compost, its use efficiency was quite high.

The physico-chemical properties of soil were also improved by the application of compost (Table 2.11). In addition to the depth of top soil, total carbon, and the extractable Si in the soil (extracted with 0.2 N HCl) was significantly increased compared to original soil. This implies that Si in compost also increases the amount of soluble Si in soil. By contrast, in the non-composted plots, the available Si in soil decreased significantly because of the uptake by rice.

Table 2.9
Effect of fertilizer treatments on Si contents of the straw of rice and wheat harvested in 1974

Treatment	Rice straw (SiO_2 %)		Wheat straw (SiO_2 %)	
	Composted	Non-composted	Composted	Non-composted
-NPK	15.16	14.02	12.21	8.94
N	13.18	10.33	8.57	7.22
P	13.86	11.20	11.91	8.14
K	16.51	12.83	11.76	9.32
PK	12.34	10.64	11.36	6.30
NK	12.84	11.87	8.79	5.74
NP	12.44	8.87	4.01	2.98
NPK	13.31	10.91	4.26	2.83
Average	13.71	11.33	9.11	6.43

The difference in the solubility of Si in the rice straw between the short-term and long-term experiments may be attributed to the slow release of Si from the rice straw (Figure 2.4). Although the effect of compost application on the Si-supplying capacity of soil was obvious in the long-term experiment as described above, it takes many years to utilize all Si contained in the rice straw. The slow release of Si from rice straw gave birth to the fast-release inorganic silicate fertilizers.

Table 2.10
Estimation of total amounts of nutrients taken up by rice plant during 40 years of cultivation

	Amount of nutrient mineral taken up (ton ha^{-1} $40yr.^{-1}$)		
	Composted, A	Non-composted, B	A/Bx100
SiO_2 (Si)	35.721 (16.667)	19.545 (9.120)	183
N	2.758	1.936	142
P	0.742	0.564	132
K	4.183	2.736	153
Ca	1.413	1.026	138

Table 2.11
Physico-chemical properties of the soils at the start and the end of the 40-year experiment

	Composted plots[a]	Non-composted plots[a]	Start of expt.
Depth of top soil (cm)	18.6	16.4	13.0
Apparent density	1.16	1.24	1.26
Total Carbon (%)	1.343	0.833	0.917
Total Nitrogen (%)	0.121	0.083	0.118
C/N ratio	11.1	10.0	7.8
0.2 M HCl soluble			
P (ppm)	156	153	117
K(ppm)	51	44	66
Ca (ppm)	1057	786	1638
SiO_2 (ppm)	1120	902	1062

[a]average of 8 plots.

2.2.2. Rice husk

Rice husk accounts for about 20% of the weight of paddy (unhusked rice) and up to 20% consists of SiO_2. Since the annual rice production in Japan is about 10 million tons as husked rice, 2 million tons of rice husk containing nearly 0.4 million tons of SiO_2 is produced annually. Rice husk is produced locally at the threshing spot. Therefore Si contained in the husk could be a potential Si source for paddy fields.

Table 2.12
Effect of carbonized rice husk application on the Si content of rice seedlings*

Treatment (SiO_2 g/plot)	SiO_2 content (%)
Control (0)	5.04
Silica gel (160)**	5.68
Burnt rice husk ashes (160)	6.51
Carbonized rice husk	
160	6.99
320	8.11
480	9.21

*1 plot is 3.3 m^2

** Silica gel was prepared by neutralizing Na_2SiO_3 solution with H_2SO_4 and thoroughly washed with water

Table 2.13
Effect of application of carbonized rice husk on Si content of rice seedlings at the seed beds of farmers

Carbonized rice husk	SiO_2 content (%)	
	Paddy seed bed	Upland seed bed
Applied	9.12 (13)*	7.58 (2)
Not applied	7.40 (11)	5.30 (1)

*( ) indicates the number of seed beds tested.

Ishibashi (1956) reported that carbonized rice husk is a good Si source although unprocessed rice husk is not. Carbonized rice husk was prepared by burning the rice husk slowly at as low temperature as possible. As shown in Table 2.12, application of carbonized rice husk increased greatly the Si content of rice seedlings. Si in carbonized rice husk was taken up better by rice seedlings than that in the ashes of rice husk burned according to the common practice.

Table 2.13 shows the amount of Si in rice seedlings sampled from the seed-beds of farmers. Application of carbonized rice husk increased the Si content of young rice significantly.

2.2.3. Silicate fertilizers

2.2.3.1. Calcium silicate slags

Slag was also initially applied as a liming material in Japan as in Germany (1937) and USA (1939). Soon after World War II, although Japan was having a food shortage problem, the amount of nitrogen fertilizers for crop production was only one tenth of that used in normal times. Phosphorus and potassium fertilizers that entirely relied on import were completely lacking. However, lime resources were abundant, and therefore, many trials using liming material were carried out. One of them is slag, which was found to have better effect in paddy soil than lime.

On the other hand, in the process of improvement of degraded paddy soils that were widely distributed, it was found that silicon application is effective in addition to the application of iron and base. As a source of silicon, slag that mainly consists of calcium silicate was applied in agricultural experiment stations and universities all over the country (Ohta, 1964 and many other papers and documents). Because the beneficial effect of slag was confirmed, slag was approved as a silicate fertilizer by the Ministry of Agriculture, Forestry and Fisheries of Japan in 1955. An official standard of slag as commercial silicate fertilizers was provided.

Slag is made by melting the ore containing Fe, Mn, Ni, and Cr with limestone

Table 2.14
Composition of slags used as calcium silicate fertilizer (%)

	CaO	SiO_2	MgO	MnO	Al_2O_3	Fe_2O_3
Pig iron slags	35-45	30-41	3-7	0.3-1.7	12-20	0.3-1.7
Steel mill slags	37-65	9-22	0.6-1.5	0.5-10.0	0.1-7.5	1.5-3.5
Stainless steel slags	43-48	23-28	10-15	0.5-1.0	1-4	0.5-1.2
Ferromanganese slags	33-37	25-30	4-7	5-14	4-7	0.5-1.0
Silicomanganese slags	30-45	30-40	1-4	6-10	3-12	3-10
Ferronickel slags	8-12	40-50	20-26	--	4-6	4-5
Nickel slags	17-20	40-45	20-25	--	2.5-4	14-20
Ferrochrome slags	47-53	27-32	9-12	--	9-12	0.2-0.5
Magnesium slags	50-55	29-33	8-12	0.1-0.3	1-3	2-4
Dephosphorated slags	48-50	40-45	0.2-0.4	--	1-4	0.1-0.3

and cokes in a blast furnace or electric furnace, and then cooling (by either air or water) material floated on the surface. Silicon components in the ore react with limestone, leading to separation of calcium silicate, and Fe and other metals in the ore are reduced and separated. After the metals needed are separated, slag remains as a by-product. The main components of slag are calcium silicate, Mg, Al, Fe and trace Mn, Ni, Cr, etc. are also included.

As a silicate fertilizer, slag must have more than 20% of 0.5 N HCl soluble SiO_2, more than 35% of alkali component, and the content of toxic component must be under permissible limit (0.4% Ni, 4% Cr, and 1.5% Ti). In 1987, slag having more than 10% soluble SiO_2 was also recognized as a silicate fertilizer. The composition of 10 slags used as calcium silicate fertilizers is shown in Table 2.14.

2.2.3.2. Fused magnesium phosphate

Fused magnesium phosphate fertilizer was manufactured in Japan from 1950. It is made by melting phosphate rock with serpentinite and ground after quick cooling. It contains P, Mg, Ca and Si. Fused magnesium phosphate appeared as a furnace phosphate fertilizer but contains 16~26% soluble SiO_2. Fertilizers containing more than 20% soluble SiO_2 are recognized as silicate fertilizer.

2.2.3.3. Potassium silicate fertilizer

Potassium silicate appeared as a slow-releasing potassium fertilizer in 1978. Fly ash that is produced from coal power plant is used as silicate material. Fly ash is mixed with potassium carbonate or potassium hydroxide and magnesium hydroxide and calcined at about 900°C. The main minerals are K_2O (Al, Fe)O_3 SiO_2 and K_2O MgO SiO_2 (Ando et al., 1985, 1990; Izumi et al., 1995). According to the official standard of commercial fertilizer, the fertilizer must have more than 20% K_2O of citrate-soluble potassium, 25% SiO_2 of 0.5 M HCl-soluble silicate, 3.0% MgO of citrate-soluble magnesium, and less than 3% of non-reactive water-soluble potassium. Potassium in the potassium silicate fertilizer is slowly released, and it was found that silicate in this fertilizer increases the resistance of rice to diseases and insect pests. At present (2000), the demand of this fertilizer is 50,000 tons, and more than 90% is used for rice.

In 1986, a liquid potassium silicate, which is guaranteed by 12% of water-soluble SiO_2 and 6.0% of water-soluble potassium, appeared as a readily available silicate fertilizer. It was produced by diluting potassium silicate and potassium carbonate in water.

2.2.3.4. Porous hydrate calcium silicate

Porous hydrate calcium silicate (tobamolite) used for light wall material in construction is produced from quick lime, quartz and cement, which are reacted under 180°C, 10 atm pressure. Because of the strict standards for wall material, a high percentage of nonstandardized product resulted. This waste material was used as a fertilizer. In 1993, "a light porous cement powder fertilizer" was recognized as a new silicate fertilizer and the standard for this fertilizer was decided. This fertilizer must have more than 15% of 0.5 N HCl soluble SiO_2 and more than 15% of base component.

2.2.3.5. Silica gel

Rice seedlings need sufficient Si to be taken up during seedling growth. However, the present commercial silicate fertilizers are not suitable for use in nursery bed because they contain alkali components that raise the pH and weaken the resistance to diseases. Silica gel is a possible Si source that does not increase soil pH. Silica gel is usually used as a desiccating agent and is made by neutralizing water glass, gelling, and finally dehydrating.

For potential application of silica gel in nursery beds of rice seedlings, in 1999, official standard for this fertilizer was decided. Different from other silicate fertilizers, silica gel is not dissolved in hydrochloric acid, and must have more than 80% of 0.5 N sodium hydroxide soluble SiO_2.

2.2.4. Estimation of available Si in silicate fertilizers

In an official method for quantitative analysis of slag, Si is extracted with 0.5 N HCl at a ratio of 1:150 (slag powder : 0.5 N HCl) with shaking. However, Si extracted by this method sometimes does not correlate with Si taken up by rice. Takahashi (1981) prepared 11 kinds of slag with different chemical compositions and different cooling methods, and compared the Si uptake of rice from slag-applied soils in pot experiment. As shown in Table 2.15, the percentage of Si taken up from the slag varied widely with the kind of slag, ranging from 73.2% (slag from phosphorus production) to 25.8% (ferronickel slag). The absorption percentage of Si by rice is lower in acidic slag with less than 1 molar ratio of CaO to SiO_2 (e.g. silicomanganese slag and ferronickel slag). Slag with coarse particles showed a lower percentage of Si uptake than those with fine particles and cooled slowly in air (Takahashi, 1981).

Since particle size and molar ratio of Ca to Si of the slag were not taken into account, the official method using 0.5 N HCl would not reflect the availability of Si for rice plants. Further investigations revealed that extraction with 4% ammonium citrate at pH 4.5 is suitable for evaluation of the amount of Si available in the slag for rice plants. The amount of Si extracted by this method is relatively low in an acidic slag and significantly decreased as the molar ratio of Ca to Si decreased. Furthermore, the solubility was larger in a slag with a fine particle size and was correlated with Si uptake by rice. However, the problem is the slag from the phosphorus production, which is highly available to rice, but has low solubility by this method.

Table 2.15
The availability of 0.5 M HCl- soluble SiO_2 for rice plant among 11 kinds of slags

Kinds of slags (cooling method)[a]	0.5 M HCl soluble SiO_2 (%)	CaO (%)	MgO (%)	S-SiO_2 /T-SiO_2 (%)	Availability of S-SiO_2 by rice plant (%)[b]	Average particle size (μm)
Slag from phosphorus production (B)	41.2	47.5	0.65	99	73.2	730
Converter slag (A)	17.7	38.8	7.3	94	55.0	379
Slag from pig iron manufacture						
(A)-1	30.1	41.0	6.5	95	50.8	234
(A)-2	31.0	43.1	5.9	99	45.0	622
(B)-3	33.8	44.4	6.7	100	37.3	714
(B)-4	38.8	44.7	2.3	100	29.3	842
(B)-5	33.3	43.6	4.5	99	29.0	925
Silicomanganese slag						
(A)-1	37.3	29.0	8.2	97	42.8	158
(C)-2	38.9	24.4	5.9	100	33.7	317
(B)-3	40.4	15.8	15.1	99	28.5	528
Ferronickel slag (B)	46.6	16.9	26.0	100	25.8	230

[a](A) slowly cooled in air; (B) cooled and crushed by water jet; (C) cooled in water
[b]at the maturity stage

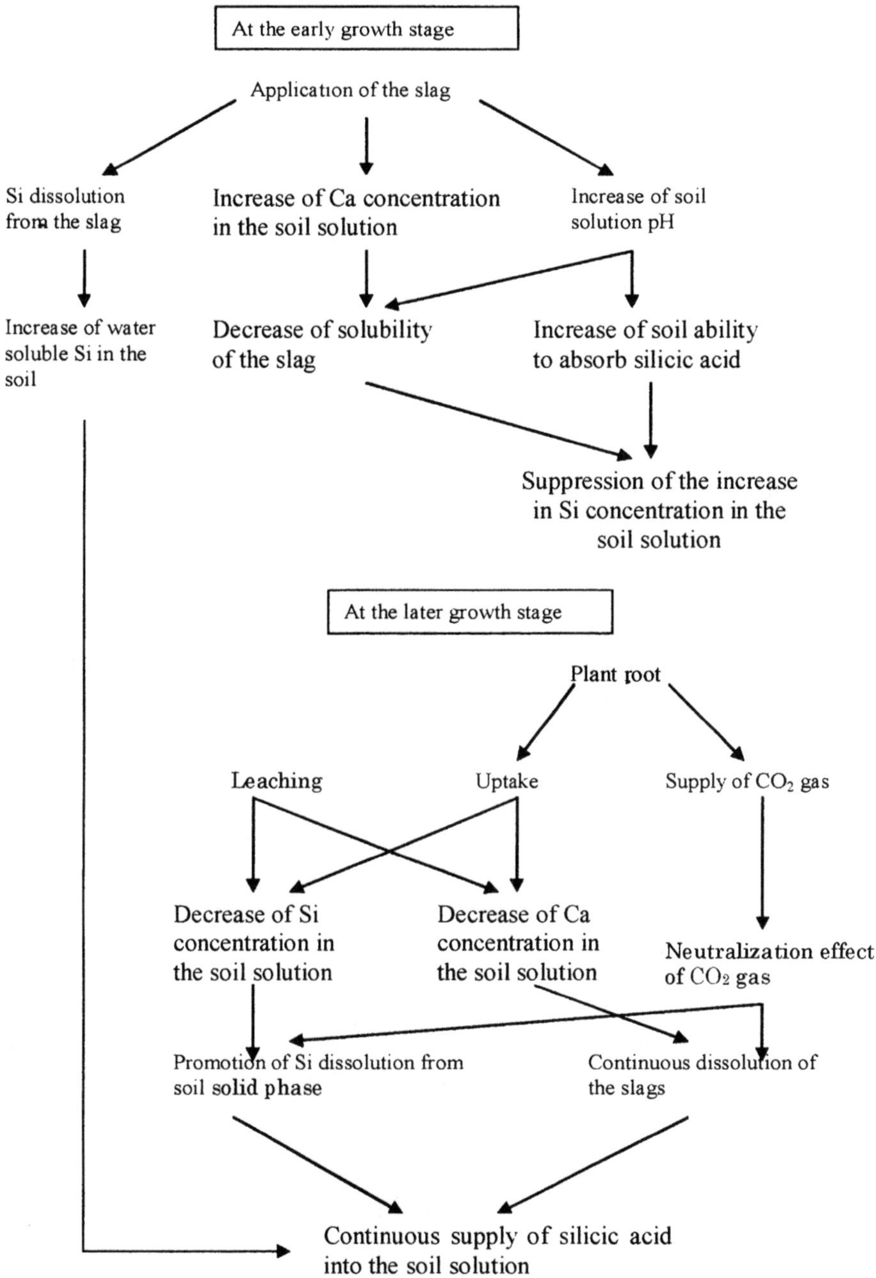

Figure 2.7. Scheme describing the possible reactions in a paddy soil with slag applied.

Recently, Kato and Owa (1996a) showed that the solubility of the slag in acid solution with a pH lower than 6 was higher than that in a solution with pH 6 or 7 and that this pH effect varied with the type of slag. These results suggest that a solution with a pH range between 6 and 7 should be used for extraction instead of an acid solution such as 0.5 M HCl when estimating the solubility of the slag in paddy soils where the soil pH is nearly neutral.

Kato and Owa (1996b, 1997a) investigated the dissolution process of slags in paddy field under rice cultivation. They found that the increase in the soil solution pH caused by the dissolution of Ca and Mg from the slag and/or the development of soil reduction, was suppressed by the neutralizing effect of CO_2 gas released by the respiration of the plant roots and the activity of the microorganisms present in the rhizosphere, hence enhancing the solubility of the slag (Figure 2.7). These neutralization effects should be taken into account for the estimation of Si availability in the slag.

Kato et al. (1997) tried to determine "active Si" in soil supplied with slag using stable isotope ^{30}Si. They found that ^{30}Si-labelled silicic acid added to the soil was diluted not only by silicic acid in soil solution, but also by silicic acid desorbed from the soil solid phase. They calculated the Si amount which takes part in isotopic dilution within 60 minutes based on ^{30}Si concentration in soil solution. This amount was named as D_{60}-value and considered as a parameter of active Si. The content of active Si was higher in the soil that was supplied slag with a higher alkali content (Figure 2.8). There was a positive correlation between Si uptake by rice and the content of active Si (Figure 2.9).

Based on these findings on the dissolution process of the slag in paddy field, Kato and Owa (1997b) proposed a new extraction method for the evaluation of the availability of Si in the slag. They dissolved the slag in water with the addition of a weakly acidic cation exchange resin (H^+ form), and examined the effects of the slag/water ratio, the amount of resin, and the temperature on the Si dissolution from the slag to determine adequate extraction conditions. The pH of the extraction was well controlled between 6 and 7 during the extraction, and the Si dissolution from the slag was enhanced by the addition of resin.

Figure 2.10 shows the relationship between the percentage of Si taken up by the rice plants from the 20 kinds of slag and the percentage of Si dissolved in 0.5 M HCl or acetate buffer solution. More than 70% of Si in the slag was dissolved in 0.5 M HCl, while the percentage of Si taken up by rice plant varied widely from 3.4 to 64.4%. The percentage of Si dissolved in the acetate buffer solution ranged from 32.8 to 95.3%, which was higher than the range of the Si recovery rate. Therefore, it is difficult to estimate the amount of Si available to the plant in the slags by these extraction methods.

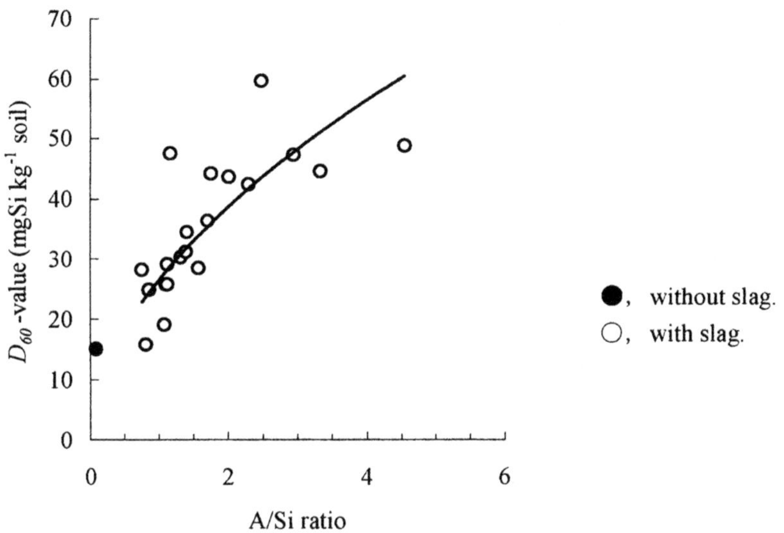

Figure 2.8. Relationship between the A/Si ratio of the slag and the amount of Si in isotopic dilution within 1 h (D_{60}-value).

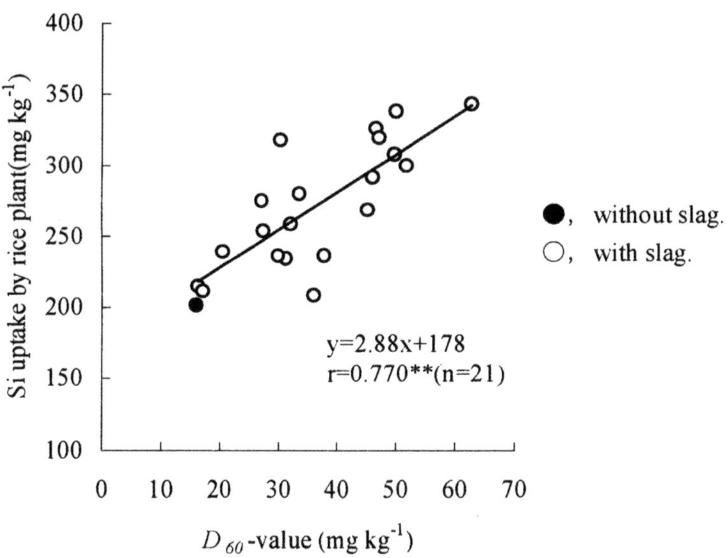

Figure 2.9. Relationship between the Si taken up by rice plant and the D_{60}-value.

On the other hand, a better relationship was observed between the percentage of Si taken up by the rice plants, and that dissolved in water from the slag (Figure 2.11). The percentage of Si dissolved either in the presence or absence of resin was higher in the magnesium slag and convertor slag that showed a high availability of Si.

Even in the extract without resin, the percentage of Si dissolved was positively correlated with that taken up by the rice plants. However, in this case, the percentage of Si dissolved was much lower than the latter. When the slag was dissolved with resin at 25°C, the correlation coefficient and the regression coefficient were slightly lower than those at 35°C, however, for the convenience of temperature control, extraction at 25°C was adopted.

Thus, the new extraction method using a cation exchange resin was established as follows: Put 0.2 g slag and 0.5 g of a weakly acidic cation exchange resin in the H^+ form (Amberlite IRC-50) into a 500 ml volume plastic bottle. After adding 400 ml of distilled water, immediately shake the bottle for a while by hand and then with a reciprocal shaker (100 rpm) at 25°C for 96 h. After filtration, determine the Si concentration in the filtrate colorimetrically.

The percentage of Si absorbed by rice plants from slag fertilizers was reported to be 20 to 50% in rice. Sumida and Ohyama (1991) investigated the percentage of Si absorbed from organic and inorganic Si fertilizers in cold area and found that the percentage was about 30% from electric furnace slag, 6%

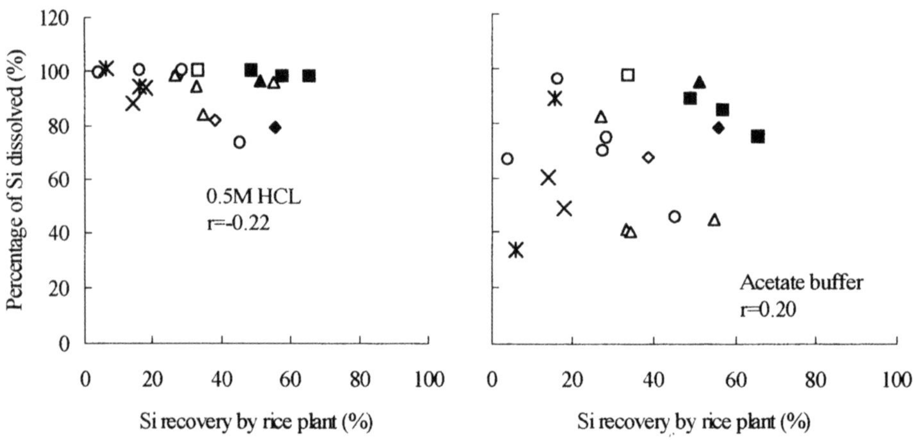

Figure 2.10. Comparison between the Si recovery rate by rice plant and the solubility of the slags in the acid solutions. ○, blast furnace slag; △, silico-manganese slag; ×, ferronickel slag; ▲, phosphorus slag; *, ordinary steel slag; ◇, stainless steel slag; □, ferrochrome slag; ◆, magnesium slag; ■, convertor slag.

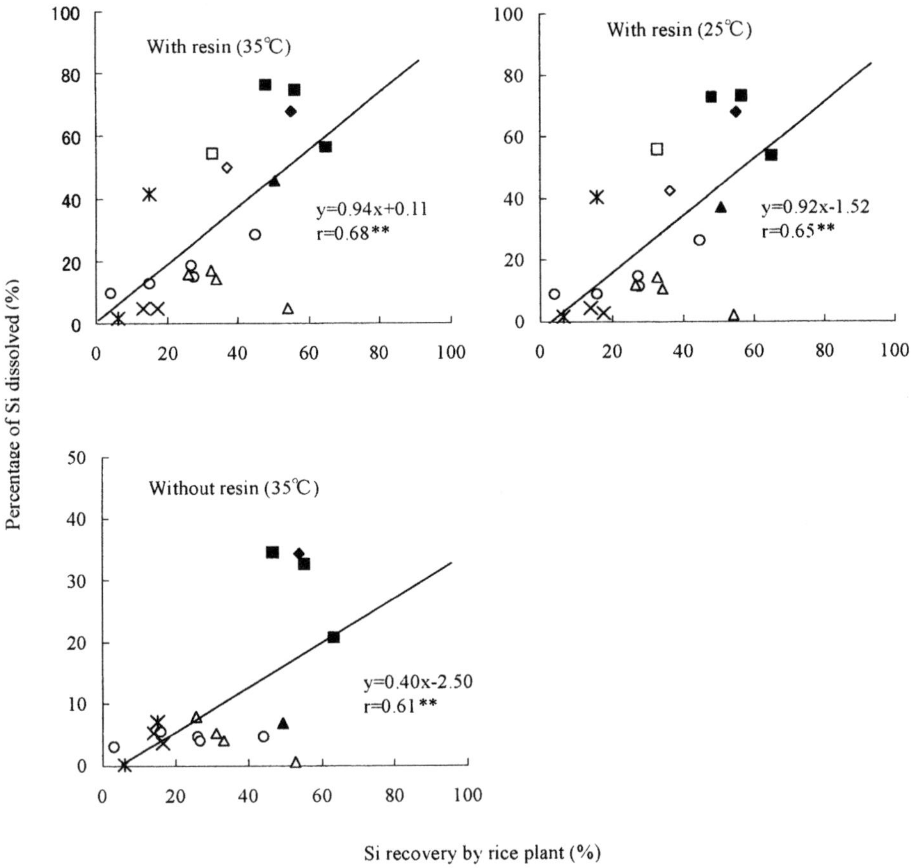

Figure 2.11. Comparison between the rate of Si absorbed by rice plants and the solubility of the slags in water with or without the addition of resin. ○, blast furnace slag; △, silico-manganese slag; ×, ferronickel slag; ▲, phosphorus slag; *, ordinary steel slag; ◇, stainless steel slag; □, ferrochrome slag; ◆, magnesium slag; ■, convertor slag. $^{**}p<0.01$.

from rice straw, and 3% from rice straw compost. In addition, although Si supply from slag was recognized at the early stage of rice growth, about 70% of available Si was supplied during the period from panicle formation stage to full heading stage. Silicon supply after heading stage may not be expected. However, as the residual effect of continuous application of slag was observed for several years after stopping the application of slag, it is suggested that slag contains readily released fraction and slowly released fraction of Si.

The percentage of Si absorbed by the rice plants from the slag was conventionally calculated based on deduction. However, Owa and Kato (1998) used calcium silicate labeled with the stable isotope ^{30}Si and found that the percentage of Si absorbed by rice was 38% in mineral acid soil and 45% in gray lowland soil. They also found that 40% of Si in rice straw in mineral acid soil and 30% in gray lowland soil resulted from calcium silicate applied. From these results, it is clear that the percentage of Si absorbed from calcium silicate differs with the soil. Gray lowland soil has a higher Si-supplying capacity and the application of calcium silicate is more effective in it than in mineral acid soil (Kato, 1998).

Chapter 3

Silicon in soil

3.1. BEHAVIOR OF SILICON IN PADDY SOIL

Behavior of silicon in paddy soil is affected by draining of flooded water, degree of reduction, pH and temperature of soil and so on.

By draining of flooded water, solubilized Si in paddy soil is eluviated together with K, Ca, Mg, etc., and with the progress of soil reduction, solubilized Fe and Mn are eluviated, leading to eluviation of solubilized Si. These are the processes happening in the degraded paddy soils.

The amount of solubilized Si increases with soil reduction, suggesting that Fe plays an important role in solubilizing Si. Although Imaizumi and Yoshida (1958) reported that part of Fe is bound to Si, it is speculated that Si is mainly bound to Al from the quantitative relationship between Si, Fe, and Al solubilized. This implies that Fe is present as a coating membrane of Si-Al complex particles, and the reactive surface of these particles increases with reduction of soil, resulting in increased solubilization of Si.

In the paddy field, soils must supply a sufficient amount of Si to rice plants. Although Si is a major component of soil, the solubility of Si compounds is usually extremely low, and a large amount of water would be necessary to solubilize a sufficient amount of Si.

Takahashi (1974) investigated the effect of soil moisture conditions on the availability of soil Si for rice plants. The soil was kept at three levels of soil moisture: upland (50% of maximum water-holding capacity (MWHC), saturated (100% of MWHC), and submerged (150% of MWHC) conditions with distilled water. After 1 month of cultivation on these soils, the amount of Si taken up by the rice plant and that of acetate buffer-soluble Si, Fe, and Al in the soils were determined. As shown in Table 3.1, the amount of Si taken up by the rice plant was the largest in the soil submerged, followed by that saturated with water and that under upland conditions, in this order, suggesting that Si uptake increases as the soil moisture is increased. The high rate of Si uptake under water-saturated and submerged conditions may be attributed to increased solubilization of Si by a large amount of water, increased diffusion of solubilized

Table 3.1
Effect of soil moisture on Si uptake by rice seedlings

Soil moisture conditions	Total amount of supplied water[4] (ml pot^{-1})	Dry weight of the top (mg pot^{-1})	Amount of SiO_2 in the top (mg pot^{-1})	Amount of acetate buffer-soluble Si, Al, Fe in the soil after harvest (mg pot^{-1})		
				SiO_2	Al_2O_3	Fe_2O_3
Upland[1]	3050	1115	40	71	312	136
Saturated[2]	4225	1240	64	60	308	161
Submerged[3]	4500	1295	78	63	310	192

[1]50% maximum water-holding capacity
[2]100% of MWHC
[3]150% of MWHC
[4]during experiment (30 days)

Si to the roots, and increased solubility of soil Si due to reduction.

On the other hand, the acetate buffer-extractable Si in the soil was less under the saturated and submerged conditions than that under the upland condition, which is in agreement with higher uptake by rice plants under saturated and submerged conditions. The amount of extractable Al was not influenced by soil moisture (Table 3.1), while that of extractable Fe was significantly increased under saturated and submerged conditions, suggesting that Fe is solubilized by soil reduction.

The amount of soluble Si in soil is an important parameter in the process of paddy soil formation and degradation. However, the research work in polder showed that solubilization of soil Si is also related to soil pH in addition to solubilization of Al and Fe. In the Kojima area of Okayama prefecture, there

Table 3.2
Changes in SiO_2 content and yield of rice plant cultivated in the Kojima polder paddy field with the lapse of time (years) after drainage

Years after drainage	Soluble SiO_2 (mg $100g^{-1}$ top soil)[a]	Yield of rice (kg $10a^{-1}$)	SiO_2 in stem and leaves (%)	SiO_2 in panicles (%)
2	27.3	600	14.1	3.7
60	13.7	550	13.1	3.3
150	9.1	540	12.5	2.2
252	8.4	480	8.8	2.5

[a] pH4 acetate buffer-soluble SiO_2

are polder paddy fields drained 1 to 324 years ago. Miyake and Yoneda (1976a, b) investigated the correlation of soil pH and the amount of soluble Si. They found that the amount of acetate buffer soluble Si in soil, rice yield, and Si uptake by rice plants decreased as the years elapsed after drainage (Table 3.2). The pattern of the changes in the amount of soluble Si was similar to that in soil pH (Figure 3.1). This suggests that soil silicate is first decomposed by acid to soluble Si, and then taken up by rice plants or eluviated. Furthermore, soluble Si is converted to insoluble silica with the decrease in soil pH.

They further investigated the effect of rice cultivation on soil properties in polder sub-soil without Si eluviation using a lysimeter. During the four years from 1968 to 1972, the soil was acidified to around pH 4, and soluble Si in the soil decreased with time in both cultivated and non-cultivated soil from the second year of the trial. However, there was no significant difference in the amount of water-soluble Si between cultivated and non-cultivated soil, suggesting that the soil keeps the concentration of Si in soil solution at a constant level even after it was taken up by rice plants.

The concentration of Si in the percolating water increased as the pH decreased rapidly. The amount of Si eluviated from rice-cultivated soil was 1/2 to 1/4 of that from non-cultivated soil. However, compared with Si uptake by rice plants, the amount of Si eluviated was extremely low; being 1/16 at the first

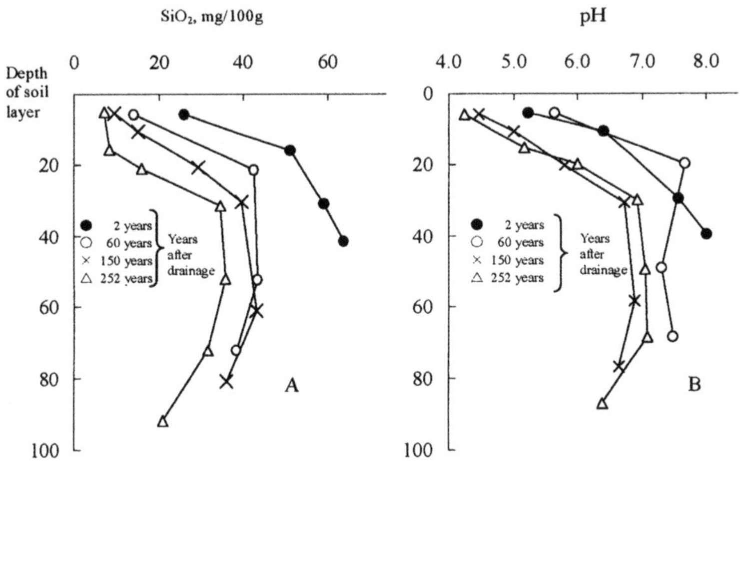

Figure 3.1. Changes in soluble SiO_2 (A) and pH (B) in Kojima polder paddy field with the lapse of time (years) after drainage.

Table 3.3
Effect of Si on the content of active Al in soil

Treatment	Content of active Al (mg/100 g soil)	
	Without compost applied	With compost applied
Control (irrigated with distilled water)	22.0	10.8
Si-enriched irrigation water	19.9	8.5
Rice straw	19.3	7.6
Slag	13.0	8.3

year and 1/30 at the second year. It was also found that more Si was eluviated during the summer season (flooded condition) than during the winter season. This may be because the solubility of Si was increased at a higher temperature and in a flooded condition.

Silicon in the irrigation water and silicate fertilizers may react with active aluminum, leading to changes in soil properties after a long period. Because volcanic ash soils distributed widely in Japan are rich in active aluminum, attempts to improve such volcanic ash soils by applying Si have been made (Onikura, 1959).

The effect of Si on active Al was also investigated by Takahashi et al. (1986). They grew rice plants in a pot filled with soils with or without compost successively applied. The content of acetate buffer soluble Si was 21.2 and 4.8 mg SiO_2/100 g in the soil with and without successive application of compost, respectively. Rice straw (0.5% of soil), or slag (Si amount corresponding to Si in the rice straw), or Si-enriched irrigation water (containing additional 14 ppm SiO_2 as silicic acid) was applied to these soils. At the end of the experiment, the content of 1N acetate buffer-soluble Al in soil was determined as a parameter of active Al. The content of active Al in the soil without compost applied was double that in the soil successively applied compost (Table 3.3). Furthermore, in both soils, the content of active Al was decreased by application of rice straw, slag, and Si-enriched irrigation water compared with the control soils irrigated with distilled water. These results indicate that Si does play an important role in reducing the amount of active Al in soil.

3.2. ESTIMATING THE SILICON-SUPPLYING CAPACITY OF PADDY SOILS

After the introduction of silicate fertilizers (slags mainly containing calcium silicate), it became necessary to establish the criteria for the applying silicate fertilizer in paddy fields.

3.2.1. Measuring acetate buffer-soluble silicon (Acetate buffer method)

Imaizumi and Yoshida (1958) working at the former National Institute for Agricultural Sciences made an intensive investigation on the Si-supplying capacity of paddy soils and Si requirement of rice plants as described below. They collected seven kinds of soil from various areas of Japan and extracted Si using various solvents (hot hydrochloric acid, ammonium oxalate buffer at pH 3.0, 2% sodium carbonate, 0.002 N sulfuric acid at pH 3.0, saturated carbonic acid water at pH 3.8, and sodium acetate buffer at pH 4.0). They grew rice plants in a pot filled with different soils. They found that rice plants take up Si from the dilute acid-soluble fraction of soil Si rather than the dilute alkali-soluble fraction (2% Na_2CO_3), suggesting that Si available for rice plants is a silicate type, and not silica type.

Among the acidic solvents examined, saturated carbonic acid water (pH 3.8) was most fitted to estimate the Si uptake by rice plants, and the amount of SiO_2 extracted with this solvent correlated well with that extracted with acetate buffer (pH 4.0) (Figure 3.2). Because it is easy to prepare, and keep the pH stable during the extracting process, acetate buffer was adopted as a solvent to determine the amount of soluble Si in soil thereafter.

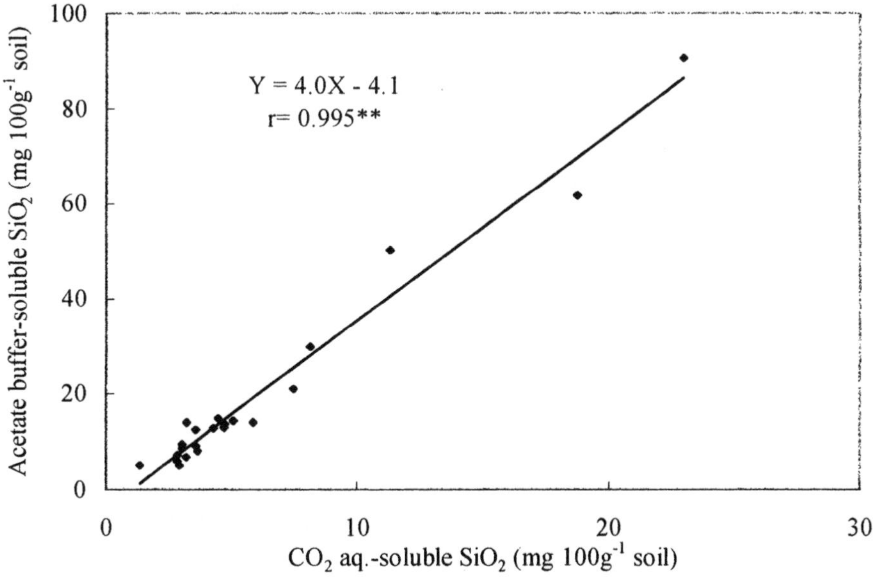

Figure 3.2. The relationship between CO_2 aq.-soluble SiO_2 and acetate buffer-soluble SiO_2.

To examine the usefulness of the acetate buffer method for determining the availability of Si for rice, pot experiments were conducted with two different soils; one uncultivated alluvial soil from Arakawa and the other degraded paddy soil from Shiga. Silicon uptake by rice plants at various time points was compared with the amount of change in the amount of Si extractable with acetate buffer. Silicon uptake by rice plants was higher in Arakawa soil than that in Shiga soil (Figure 3.3), which was well correlated with the amount of extractable Si in the soil (Figure 3.4).

After examining the temperature, the time of extraction and ratio of soil to solvent, the protocol for determining the acetate buffer-soluble Si was decided as follows. Acetate buffer is prepared by dissolving 49.2 ml acetic acid and 14.8 g anhydrous sodium acetate in 1 liter water. This solution is adjusted to pH 4.0 using acetic acid or sodium acetate solution. Ten grams of air-dried soil is extracted with 100 ml acetate buffer in a flask, and the mixture is shaken occasionally for 5 h at 40°C. After filtration with dry filter paper, the concentration of Si in the filtrate is determined by the colorimetric molybdenum blue method.

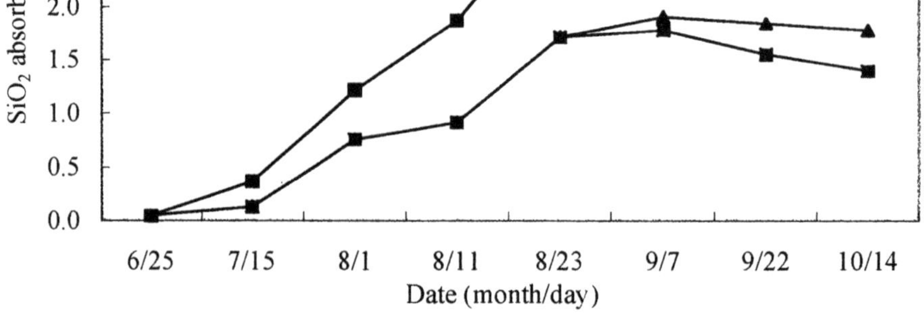

Figure 3.3. Amount of SiO_2 absorbed at various growth stages.

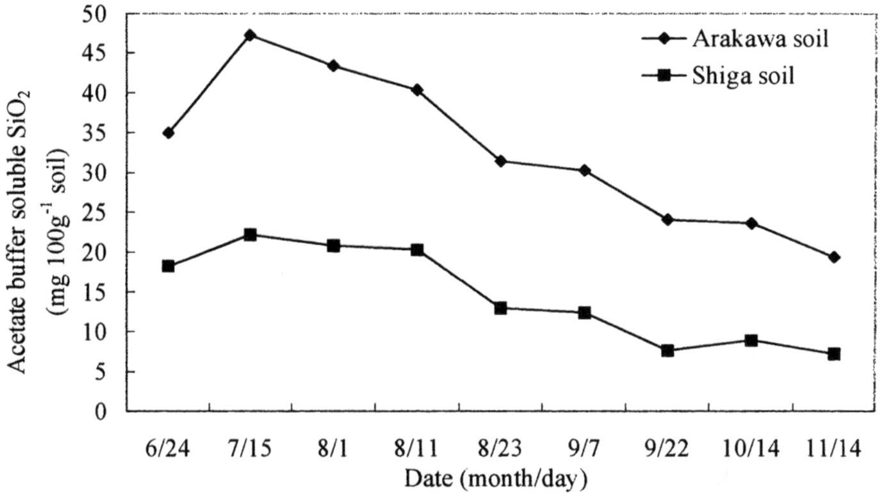

Figure 3.4. Changes in acetate buffer-soluble silicon in soils during the growth period of rice plant.

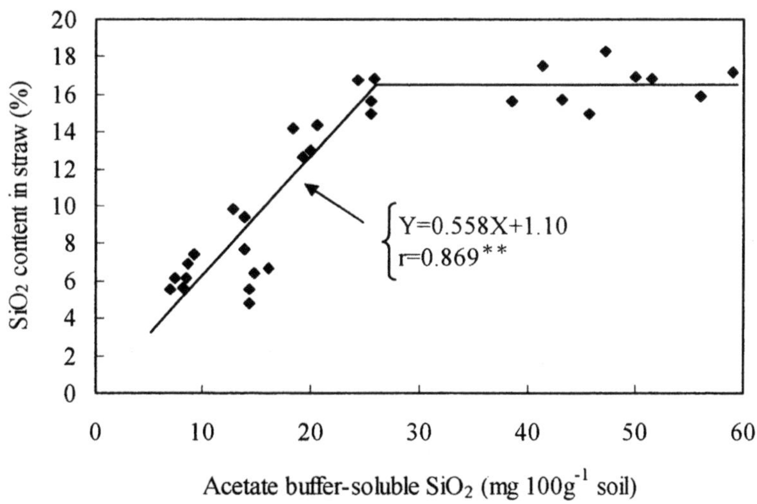

Figure 3.5. Relationship between acetate buffer-soluble SiO_2 in soil and SiO_2 content in straw.

Rice growth in the field is affected by nutrients in soils, climate and other conditions in addition to Si. The amount of Si uptake is also affected by dry matter production, while the content of Si is less affected by dry matter production. Therefore, the content of Si in the rice straw is used as a parameter of soil Si available for rice plants.

Figure 3.5 shows the relationship between the amount of Si extracted with acetate buffer and the Si content of rice straw collected from 30 fields in Tochigi and Nagano prefectures. There were two phases; one is that the Si content of rice straw increased with the amount of Si up to 26 mg SiO_2/100 g soil, and the other is that the Si content of rice straw was almost constant regardless of increase in the soluble Si. There was a good correlation (r=0.869) between Si content of rice straw and acetate buffer-soluble Si in soil in the former phase. These results suggest that extraction with acetate buffer is suitable for evaluation of soil Si available for rice plants.

Figure 3.6 shows the relationship between Si concentration in irrigation water and Si content in rice straw. A good correlation (0.772) was obtained, suggesting that Si in the irrigation water could also be a valuable parameter of the availability for rice plants. However, acetate buffer-soluble Si in soil is a more reliable parameter.

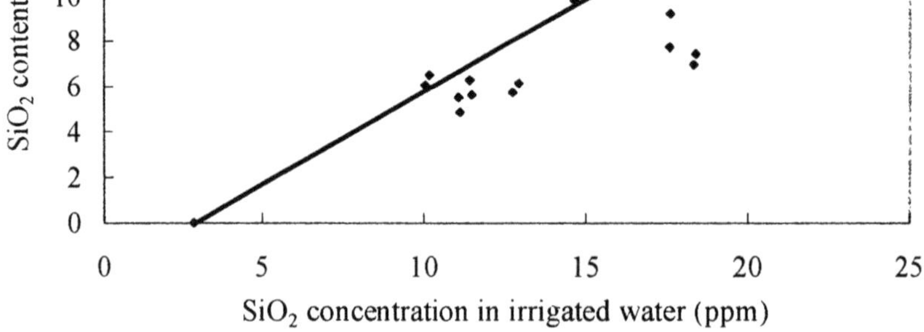

Figure 3.6. Relationship between SiO_2 concentration in irrigated water and SiO_2 content in straw.

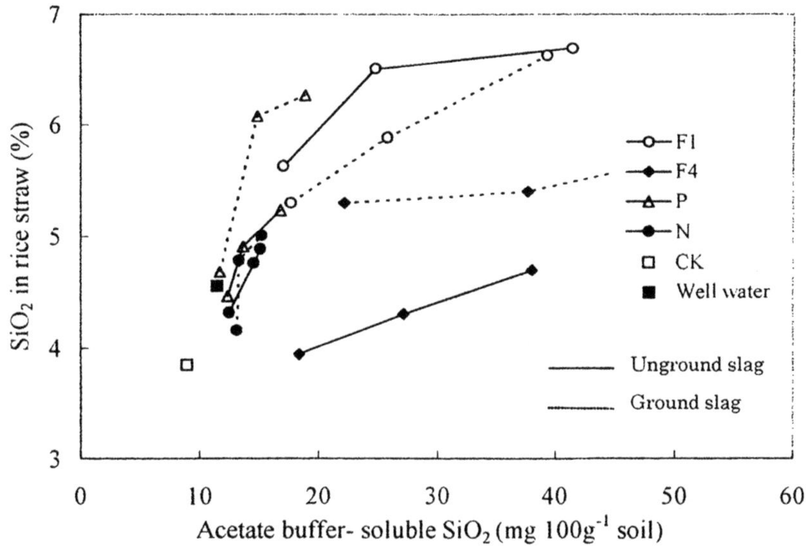

Figure 3.7. Relation between acetate buffer-soluble SiO_2 in soils amended with slags and SiO_2 content in rice straw. F1, pig iron slag (slowly cooled); F4, pig iron slag (crushed with water); P, phosphorus slag; N, ferronickel slag; CK, control.

Since 1958, the acetate buffer method had been used widely for determining soil soluble Si. However, in the late 1970's, the amount of acetate buffer-soluble Si did not reflect Si uptake by the rice plants. In some soils applied slags, extremely high soil soluble Si was detected. As shown in Figure 3.7 (Takahashi, 1981), soil soluble Si extracted by acetate buffer differed greatly with kinds of slags and soils, and was not correlated with Si content of rice straw. However, there was a good correlation between the molar ratio of Si/Al and the Si content of rice straw (Figure 3.8, Takahashi, 1981). These results suggest that acetate buffer is strong enough to extract Al-bound Si in slags which is unavailable for the rice plant. Acetate buffer method is originally developed for evaluation of availability of natural Si in soil, therefore, for evaluation of the availability of Si in the soils with slags applied, a new method is necessary to be developed.

3.2.2. Measuing Si dissolved under submerged condition (Incubation method)

Takahashi and Nonaka (1986) at former Shikoku Agricultural Experiment Station tried to develop a method for evaluation of soil soluble Si which can

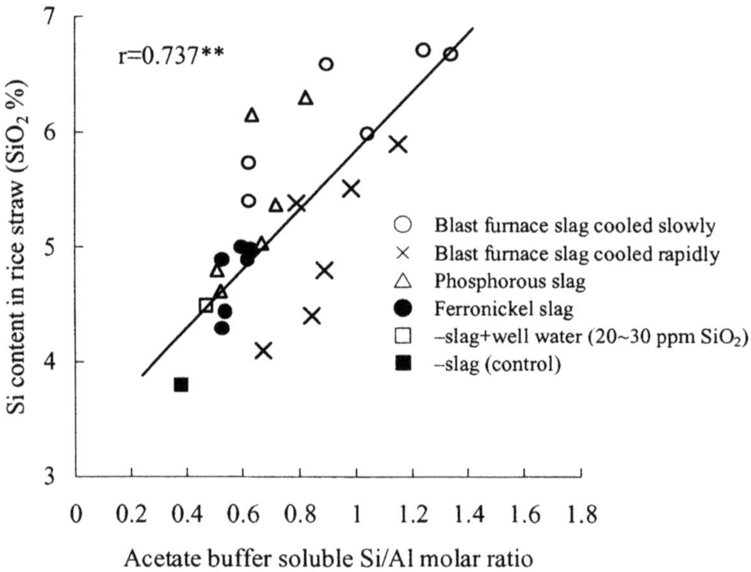

Figure 3.8. Relationship between the molar ratio of SiO_2/Al in the extract of soils amended with slag and SiO_2 content in rice straw.

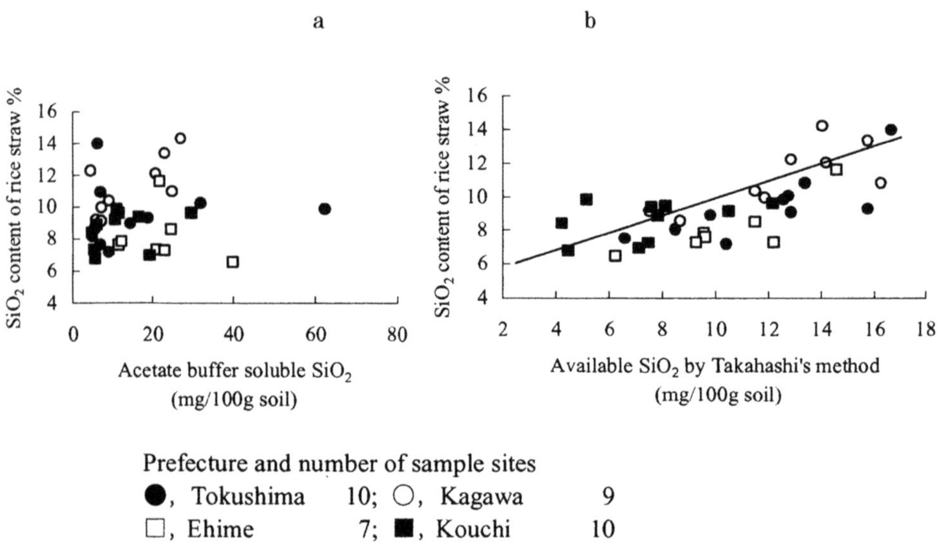

Figure 3.9. Correlation of SiO_2 content of rice straw with the amount of SiO_2 of paddy soil extracted by acetate buffer (a) and Takahashi's method (b).

reflect Si uptake by rice plants and is not affected by the application of slags.

After various examinations, they established a method which was called "Incubation under submerged condition". The procedures of this method are as follow: Taken 10 g of air-dried soil in 100 ml polyethylene bottle containing 60 ml distilled water. After shaking and degassing, the bottle is sealed and incubated at 40°C for 1 week without shaking. Finally, the Si concentration in the supernatant is determined.

In contrast to the acetate buffer method (Figure 3.9a), a good correlation was obtained between the Si content of rice straw and soil soluble Si determined by this method, irrespective of slag application (Figure 3.9b). This result suggests that this method could be used to evaluate Si supply power of soils. In a book titled "Standard Analysis Method of Soil" published in 1986, this method was adopted.

Soil soluble Si estimated by this method was higher at early stage of rice growth and gradually decreased thereafter. Because this phenomenon was not observed in non-cultivated soils, the decrease in soil soluble Si during rice growth period is attributed to active Si uptake by rice plants. However, as shown in Table 3.4, the amount of soluble Si recovered to original level around cropping season of the following year (Takahashi, 1986). Although the mechanisms for the recovery are unknown, it seems that soil has capability to keep soluble Si at constant level through a cycle of decrease and increase during a year.

3.2.3. Measuring silicon in supernatant (Supernatant method)

In the measurement of Si by the incubation method (3.2.2), there is a possibility that Si might precipitate during the incubation due to oxidation by air. To improve this point, Sumida et al. (1988) used a method named "Supernatant method" which determines Si concentration in the supernatant of highly reduced soil. Soil is put in a 50 ml test tube (30 mm D x 110 mm H), with distilled water at a ratio of 1:4 (soil to water). After shaking and

Table 3.4
Seasonable changes of water soluble SiO_2 by incubation method

	1982		1983		1984	
	June 25	Oct. 22	June 27	Oct. 22	June 4	Oct. 22
			SiO_2 (mg $100g^{-1}$ soil)			
Successively composted[a]	25.1	19.3	22.8	17.5	25.2	19.9
Non-composted	15.2	10.4	13.6	9.4	15.3	12.6

[a] composted 10t/ha for 4 years

Table 3.5
Amounts of soluble soil Si measured by various methods

Treatments of soil		pH4 acetate buffer method	Incubation method	Supernatant method
		SiO_2 mg / 100 g soil		ppm
Compost (t/ha)	0	14.5 ± 1.1	9.4 ± 0.2	10.4 ± 0.6
	20	15.8 ± 1.1	7.1 ± 0.1	14.4 ± 0.3
	40	18.2 ± 1.6	8.6 ± 0.3	19.1 ± 0.4
Rice straw		14.5 ± 1.5	9.0 ± 0.3	18.7 ± 0.6
Calcium silicate		66.9 ± 1.5	10.6 ± 0.5	19.5 ± 0.2

degassing with voltex, water is added to fill the tube to replace air space and then sealed. The tube is incubated at 30°C for four weeks and then the Si in the supernatant solution is determined by a colorimetric method.

The amounts of Si extracted from soil by various methods are compared in Table 3.5. The amount of Si extracted by incubation under a submerged condition was smaller than that extracted by other methods in either soil supplemented with compost or rice straw or calcium silicate. On the other hand, the amount of Si determined by the supernatant method was doubled by the application of calcium silicate to the soil. And the amount of Si extracted by this method increased as the application rates of compost increased, while the amount of Si extracted by incubation under a submerged condition was hardly affected by the application of compost. These results suggested that in the soils with intensively reduction caused by compost application, the amount of extractable Si is decreased by oxidation during the extraction process. Since the amount of Si in soil determined by the supernatant method was increased by the application of calcium silicate or compost or rice straw to the soil as a source of Si, and was well correlated with the amount of Si taken up by the rice plants, this method is considered to be useful to estimate the amount of Si in soil available to the rice plants.

3.2.4. Measuring easily soluble Si (Easily soluble Si method)

Sumida (1991) investigated the characteristics of Si dissolution and adsorption (SDA) of various soils with different history of fertilizer management. He tried to find out a way to predict Si supplying capacity of paddy soils from the SDA. He added silicic acid solution ranging from 0 to 100 ppm to soil (soil:water ratio 1:10) and then incubated it at 30°C for 5 days. By measurement of Si concentration in the filtrate, Si dissoluted or adsorbed was calculated. The difference between initial and final Si concentration in solutions through incubation (y ppm) was linearly proportional to the initial Si concentration in solution (x ppm) as shown in Figure 3.10 and is shown by the following equation:

$$x/a+y/b=1$$

Where, constant-a is the Si concentration of solution when neither dissolution nor adsorption occurred, and constant b is the concentration of Si dissolved from soil into distilled water. This equation was rewritten as follows

$$u/a+v/c=1,\ c=a \cdot b/(a-b)$$

Where u (ppm) is the final Si concentration in the solutions, v (mg/100g soil) is the amount of dissolution or adsorption of Si by a unit soil. Constant-c, which is extrapolated, shows the amount of Si dissolved from soil into vast water (Figure 3.11). This c value (easily soluble Si) has a better correlation with Si content of rice straw than the value obtained by supernatant method (Figure 3.12), implying that the value determined by this method could be a valuable parameter in the diagnosis of Si supply capacity of paddy soils.

3.2.5. Measuring Si dissolved in surface water (Surface water dissolution method)

The amount of Si extracted from soil by incubation under submerged method and supernatant method is only a small part of Si available for the rice plants, suggesting that these methods are a parameter of intensity of soil Si supplying

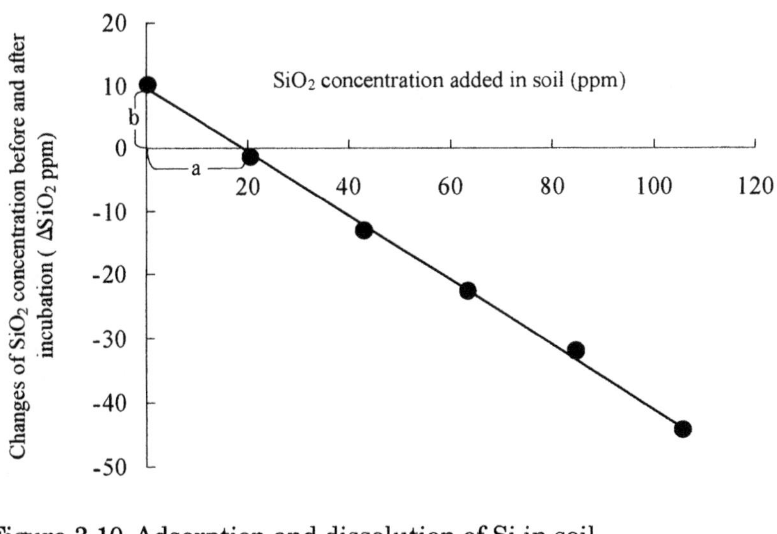

Figure 3.10. Adsorption and dissolution of Si in soil.

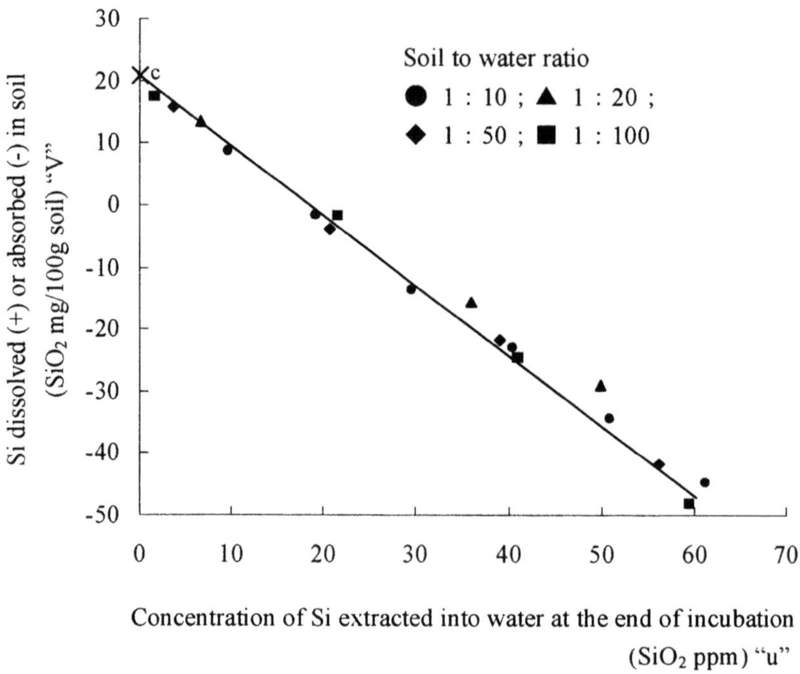

Figure 3.11. Correlation of the amount of Si dissolved or absorbed in soil with the concentration of Si extracted into water.

capacity. Kitada et al. (1992) developed a new method of which the quantity of soil Si supply was taken into account by extracting soil with deionized water several times. To field-moist soil (equal to 15 g dried soil) which had been passed 4 mm sieve in a test tube, deionized water is added at a ratio of 1:4.5 (soil: water). The tube is incubated at 30°C for 42 days under submerged condition. During the incubation, the supernatant is replaced with water every week and the Si concentration in the supernatant is determined.

A comparison was made between this method and the supernatant method as described above (Table 3.6). The amount of silicon extracted by the supernatant method was lower than that by surface water dissolution method, being one third at 42 days after the incubation. This result suggests that extraction of Si into the supernatant is inhibited by some factors in the supernatant method, while such inhibition is avoided in the surface water dissolution method. This method could be especially useful for understanding the dissolution process of Si in soil, but an incubation period of more than ten weeks is required.

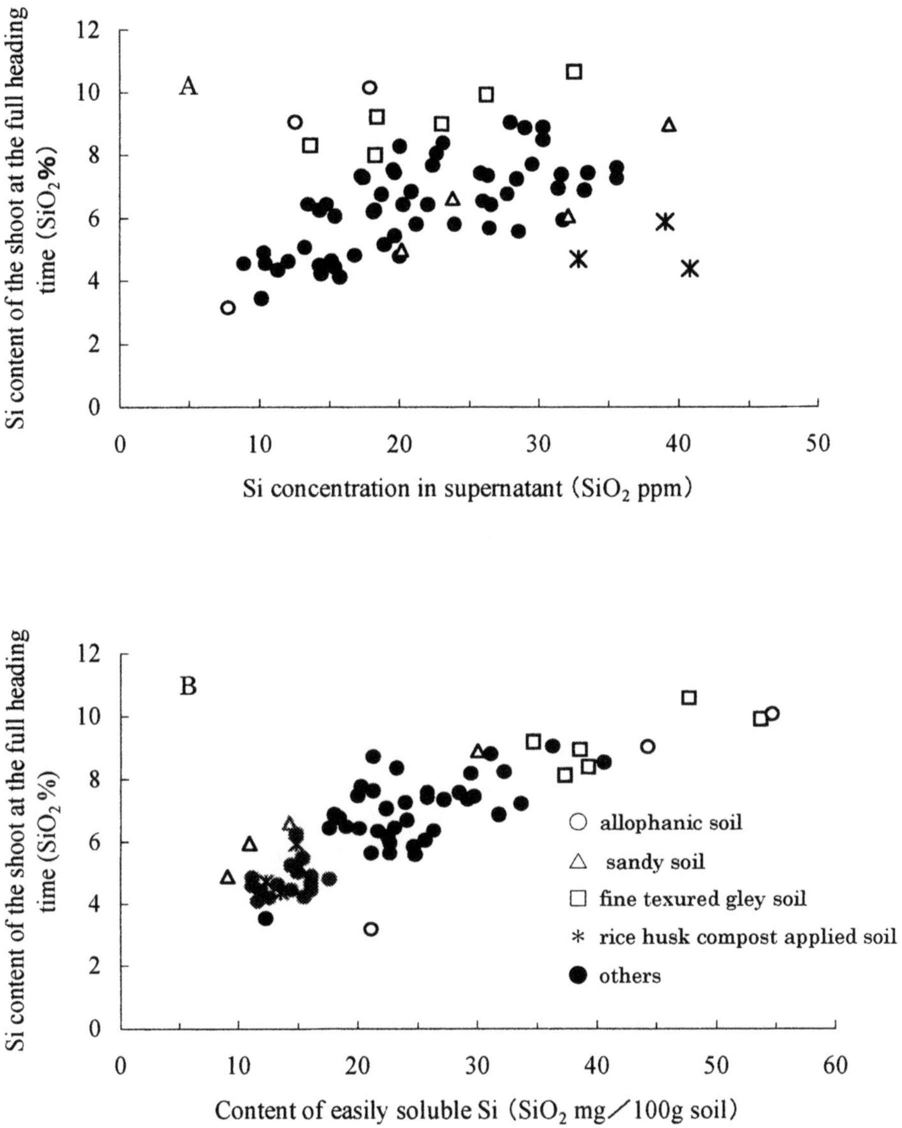

Figure 3.12. Relationship between Si content in rice shoot and the content of soluble Si determined by supernatant method (A) and easily soluble Si method (B) in various soils.

Table 3.6
Dissolution process of Si from submerged soils (gray lowland soil). Soluble Si was measured by supernatant method and surface water dissolution method

		Si content (SiO_2 mg/100 g soil)		
	Incubation days	14	28	42
Continuous cropping paddy field	Supernatant method	4	5	6
	Surface water dissolution method	8	14	19
Rotational paddy field	Supernatant method	5	6	7
	Surface water dissolution method	10	17	23

3.2.6. Measuring Si dissolved in phosphate buffer (Phosphate buffer method)

Although several methods for estimation of the amount of soil Si available for rice have been developed to improve the acetate buffer method as described above, these methods need a long time for extraction of Si. A rapid method is required for analysis of a large number of samples. Mizuochi et al. (1996) classified Si extractable from soil into two fractions. One is the exchangeable Si fraction, which is rapidly released as the Si concentration of soil solution decreased and the other is the slow-releasing fraction, which is bound to Fe and Al and solubilized only by soil reduction and chelation. As phosphate is similar in chemical form to silicate, they suggested the use of phosphate to extract exchangeable Si fraction within a short time. To 1 g soil, add 10 ml 0.02 M sodium phosphate buffer (pH 6.9-7.0), and extract Si at 40°C for 5 hours (for rapid extraction, 80°C for 30 min). The amount of Si extracted by this method showed a good correlation with the Si content in rice straw in some soils. In brown lowland soil and humic Andosol, the content of Si in rice straw was not so high while that of Si extracted by this method was high. This may be attributed to the high content of active Al in the clay fraction of these soils.

The amount of Si extractable by the Mizuochi method was very low, and was less than 10% of Si available for rice plants. Kato and Sumida (2000) improved this method by changing the phosphate buffer concentration from 0.02 M to 0.04 M, pH from 6.9 to 6.2, and extraction time from 5 hours to 24 hours at 40°C. By these improvements 3.5 times more Si was extracted from the soil compared with that by the Mizuochi method. Furthermore, a better correlation was observed between the amount of Si taken up by the rice plants and that of Si extracted from the soil by the improved phosphate buffer method, and a good correlation was observed even in brown lowland soil (Figure 3.13). However, in humic Andosol, a similar correlation was not observed.

Because a large amount of phosphate is present in the buffer used for extraction, attention should be paid to the interference of phosphate when Si in the extract is quantified. Usually, interference of phosphate could be avoided by adding tartrate before adding the reducing agent. However, the presence of a high concentration of phosphate causes formation of a phosphate-molybdate complex, resulting in insufficient molybdate to form a Si-molybdate complex. Therefore, Si in the extract will be underestimated, but this problem could be solved by increasing the amount of molybdate and tartrate.

In 2001, the book titled "Analysis method for soil, water and plant" was revised. The acetate buffer method was not adopted in this new version, while the incubation method and phosphate buffer method (improved by Kato and

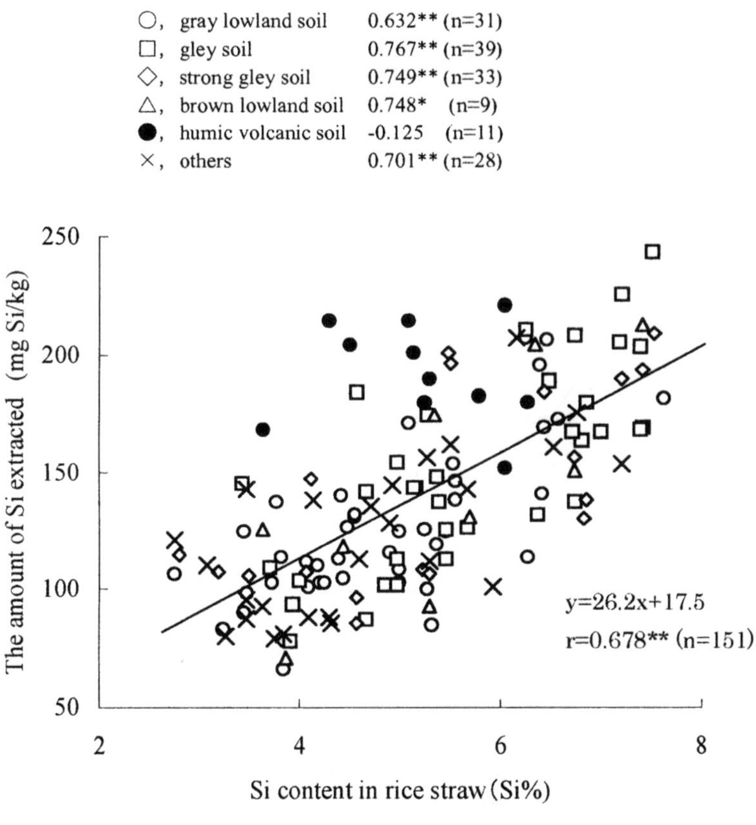

Figure 3.13. Relationship between Si content of rice straw and the amount of Si extracted from the soil with pH 6.2 phosphate buffer solution.

Sumida) were adopted. Both methods will probably be used, but one of them may be selected out in the near future.

3.3. ENVIRONMENTAL FACTORS CONTROLLING THE AVAILABILITY OF SILICON FOR RICE PLANTS IN PADDY SOILS

Sumida (1996) summarized the environmental factors controlling the availability of Si for rice plants in paddy soil (Figure 3.14). Soil Si can be divided into two parts; one is Si solubilized without development of soil reduction, and the other is Si solubilized with progress of soil reduction. The effect of soil reduction on the solubility of Si is less than that of P. Soil temperature also affects the solubility of soil Si. Si is more soluble at a high temperature. Moreover, soil has a capacity to keep the Si concentration in soil

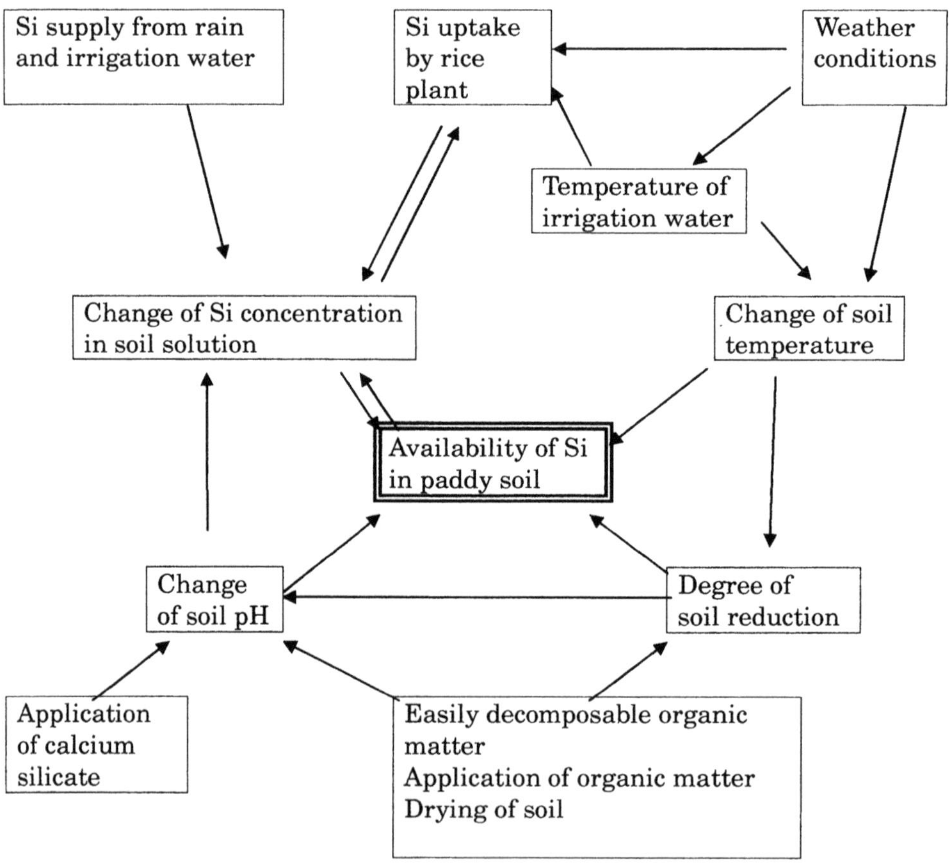

Figure 3.14. Environmental factors ruling the availability of SiO_2 in paddy soil.

solution at a constant level when the Si concentration in soil solution decreases with Si uptake by rice plants. This capacity is controlled by characteristics of Si dissolution-adsorption of soil as mentioned above.

In addition, soil pH affects the availability of soil Si. Most paddy soils in Japan are acidic. However, soon after submergence, the pH of soils approaches to neutral. As soluble Si concentration in water extract usually increases with decreasing pH, Si concentration in soil solution will decrease after submergence. All factors affecting the solubility of soil Si should be taken into account for the diagnosis of Si- supply power.

3.4. BALANCE SHEET OF SILICON IN PADDY SOIL-PAST AND PRESENT

Rice plants take up a large amount of Si from paddy soils every year. As shown in Table 3.7, the amount of SiO_2 taken up for production of 100 kg brown rice is 10-fold, 40-fold, 6-fold, and 30-fold of that of N, P, K and Ca, respectively. On the other hand, Si is supplied to the paddy soil from irrigation water, rice straw and compost, and/or silicate fertilizers. A balance sheet for Si output and input was calculated in 1987 by Yakabe at Calcium Silicate Fertilizer Association. Silicon taken up by rice plants was estimated based on planted area of rice plants of each prefecture and yield of brown rice per hectare which were reported by Ministry of Agriculture, Forestry and Fisheries, assuming that 20 kg SiO_2 was required for production of 100 kg brown rice. Silicon supply from irrigation water was estimated from the Si concentration in the main 380 rivers investigated by Kobayashi. It was calculated as a product of average Si concentration in the rivers and average irrigation amount (14.4 ton per hectare). Silicon supply from rice straw plus compost was estimated assuming that 2 tons of rice straw plus compost were applied per hectare (from investigation of application rate by Ministry of Agriculture, Forestry and Fisheries) and that they contained 5% SiO_2. Silicon supply from silicate fertilizers was calculated from the concentrations of Si in the fertilizers and the amount of silicate fertilizers consumed in each prefecture.

From these data, a balance sheet of Si revenue and expenditure for the whole country was made as shown in Table 3.8. It is clear that the revenue and

Table 3.7
Amount (kg) of nutrients taken up for the production of 100 kg brown rice estimated by Central Agricultural Experiment Station

N	P	K	Ca	SiO_2
2.1	0.5	3.3	0.7	20.0

Table 3.8
Balance sheet of SiO_2 in paddy field in 1987

Total area harvested (10^3 ha)	Total Rice Yield (10^3 t)	Total SiO_2 taken up by rice plant (10^3 t)	SiO_2 (10^3 t) supplied from				SiO_2 exhausted from soil (10^3 t)
			Silicate Fertilizer	Compost	Irrigation water	Total	
2,122.8	10,570.9	2,114.2	297.6	212.3	618.3	1,128.2	986.0

expenditure of Si is in deficit in paddy soils of whole country and the Si exhausted was as high as 0.99 million tons per year, which is equal to 460 kg SiO_2 per hectare in average. This fact suggests that the Si-supplying capacity of paddy soils is reducing year by year. From data estimated for each prefecture, the deficit of Si in Yamagata and Akita (720 kg), and Niigata (690 kg) prefectures, the main rice-producing area, is larger than in the other areas (Figure 3.15).

Silicon supplied from irrigation water accounts for 60% of total amount of Si supply. In Kyushu area, where Si is exhausted is the least (the average of 7 prefectures is 200 kg SiO_2/ha/year) (Figure 3.15), the concentration of Si in the irrigation water was 30.9 ppm, which is 50% higher than average of whole country (21.6 ppm). Therefore, the contribution of Si from the irrigation water is high in this area. However, from the experiment using a lysimeter, Si in the irrigation water was also eluviated.

Rice yield at present is double of that before the war. This implies that Si taken away from the paddy soils is also doubled. However, return of rice straw compost to the paddy soils at present decreased to one third of that before World War II. Therefore, the input and output of Si in paddy soils before the war could be estimated as follows. The average planted area of rice during 1920-1940 was 3 million hectare, and average yield was 3.0 tons per hectare. The annual application of compost was more than 20 million tons. Based on these data, SiO_2 taken up by rice plants was estimated to be 600 kg per hectare, while Si supplied from the compost and irrigation water was 330 kg and 290 kg (totally 620 kg per hectare), respectively. Therefore, the output and input of Si in the paddy soils were well-balanced before the war.

The unbalance of Si output and input in Japanese paddy soils was tentatively improved by the introduction of silicate fertilizers. However, the annual application of silicate fertilizers is decreasing year by year, from 1.3 million tons in 1968 to 0.28 million tons in 1998. Further reduction of Si fertility of paddy soils is anticipated. For example, in Yamagata prefecture, it was reported that the amount of soluble Si in soil and the Si content of rice shoot after full heading date decreased. The Si concentration in the irrigation water at present was only half to one-third of that of 40 years ago (Table 3.9, Kumagai et al., 1998).

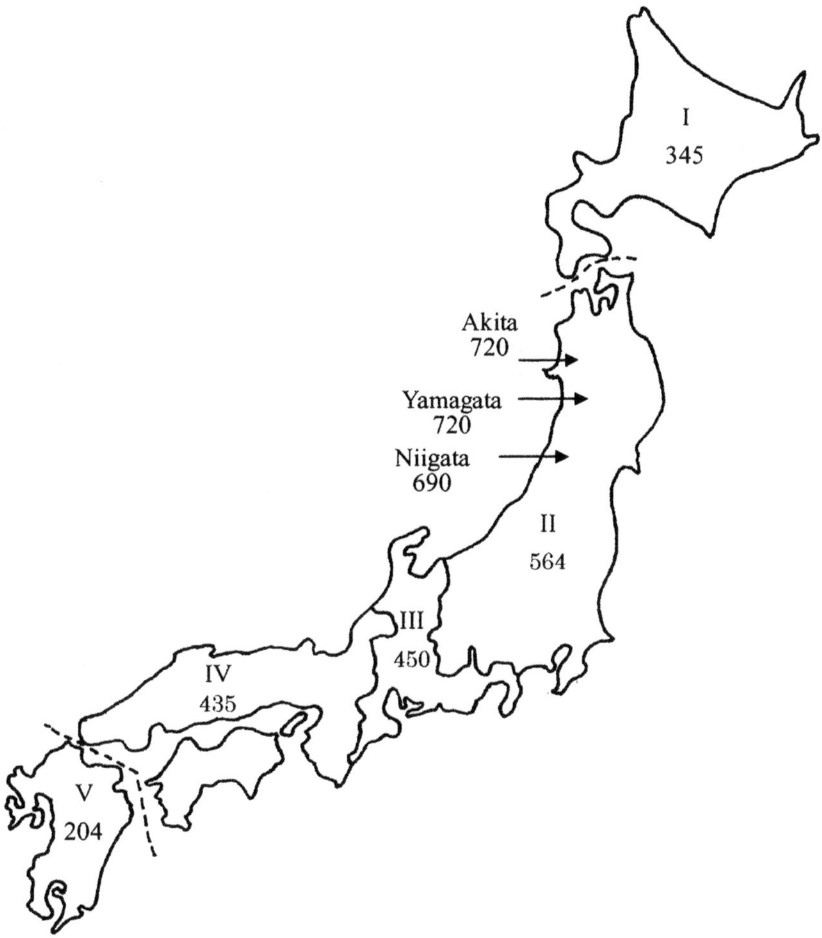

Figure 3.15. Average amount of SiO_2 exhausted from paddy field in 5 regions of Japan (kg SiO_2 /ha/year).

Table 3.9
Changes of SiO_2 concentration in irrigation water in Yamagata prefecture between 1956 and 1996. Data are mean value ± STD.

Regions of Yamagata pref.	SiO_2 (ppm)	
	1956	1996
A	19.7 ± 4.2	9.8 ± 3.0
B	22.6 ± 5.5	11.7 ± 4.5
C	30.4 ± 11.9	10.6 ± 2.6
D	19.4 ± 7.4	8.9 ± 1.7
Average of Yamagata pref.	23.9 ±9.7	10.2 ± 3.2

Furthermore, the pH in the irrigation water in this prefecture increased from 5.6 in 1956 to 7.4 in 1996. Since pH in the irrigation water has a negative correlation with the Si concentration (Figure 3.16), the decrease in Si concentration in irrigation water is considered to be caused partly by an increase in pH. After 1956, many changes have occurred in the environmental conditions such as improvement of acidic rivers, increase of dams and concrete canals, decrease of reservoirs, etc. These changes may be attributed to the changes in pH and Si concentration in the irrigation water.

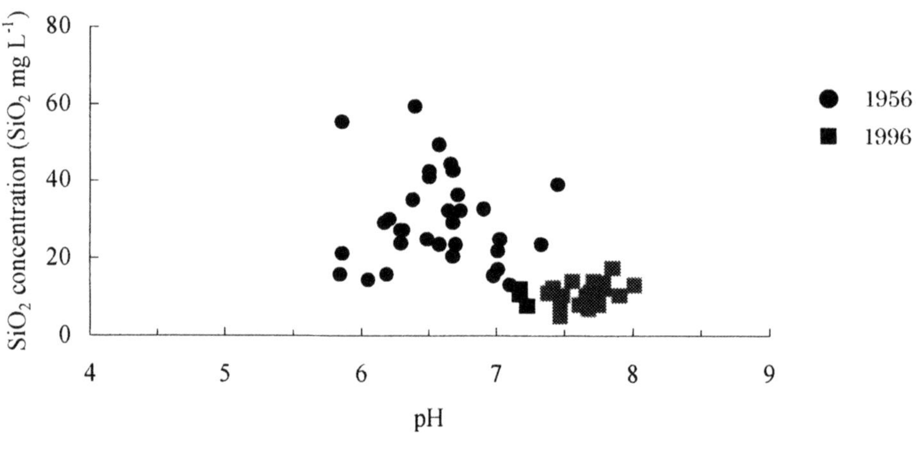

Figure 3.16. Relationship between pH and SiO_2 concentration of irrigation water (C region).

Chapter 4

Effect of silicate fertilizer application on paddy rice

4.1 CRITERIA FOR PREDICTING SILICATE FERTILIZER REQUIREMENT FOR PADDY RICE

Imaizumi and Yoshida (1958) examined the relationship between the effect of calcium silicate application on the yield and the plant available Si (pH 4 acetate buffer extractable) content in soil or the Si content in rice straw from the results of field trials conducted at agricultural experiment stations all over the country. The probability of obtaining a profitable increase in yield by the application of calcium silicate to the soil was calculated for the soils containing a definite amount of available SiO_2 or the soils on which rice straw contained a definite amount of Si. Then, the probability P was plotted against the amount of

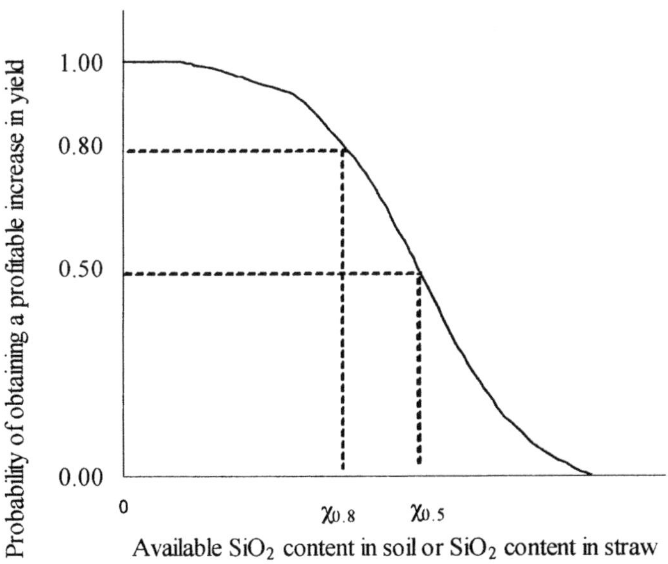

Figure 4.1. Schematic representation of P-T curves.

available Si in the soil or Si content in rice straw (Figure 4.1). This curve was named P-T curve (Probability-Test Value Curve). From this curve, we can estimate the probability of obtaining an increase in yield by applying silicate fertilizer to the soil.

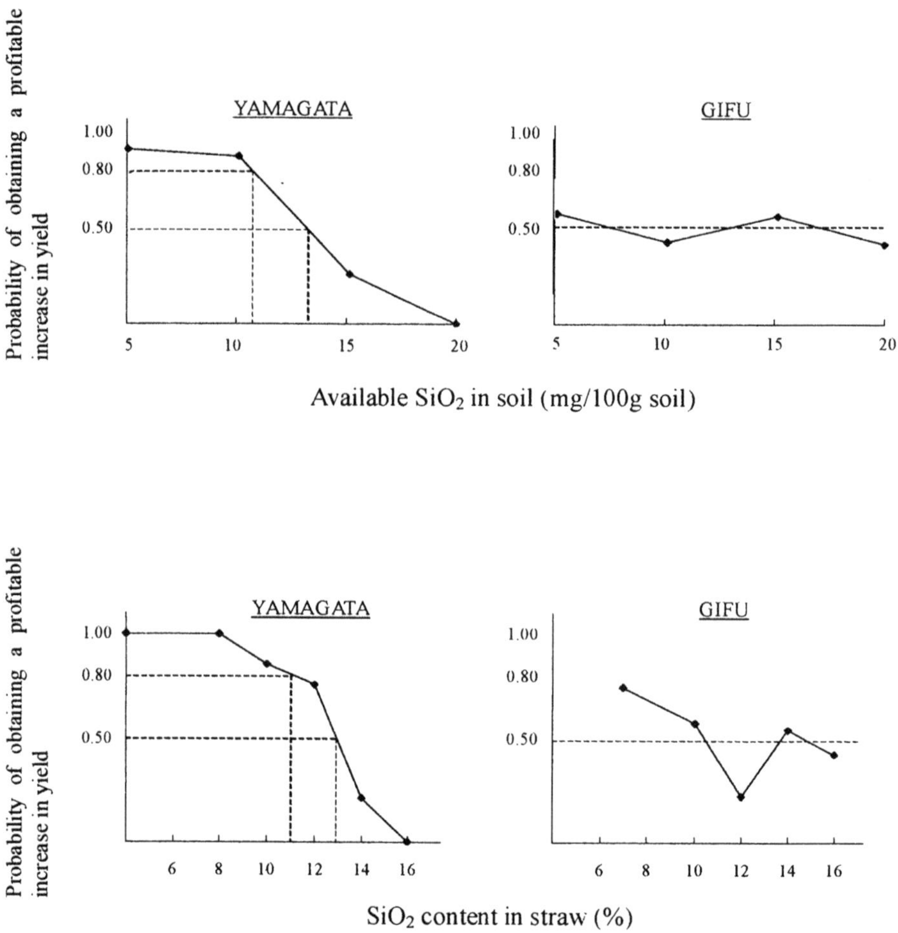

Figure 4.2. P-T curves in Yamagata and Gifu Prefectures.

The effect of fertilizer application in Yamagata and Gifu Prefectures in 1957 can be obtained from the P-T curve (Figure 4.2). Silicate fertilizer was regarded as effective when the yield increase was more than 5% or damage due to diseases or pests was alleviated (1 point), as partly effective when the yield increase was less than 5%, but total growth was good (0.5 point), and as ineffective when there was no positive effect (0 point). The P value was 0.5 point on the soil where rice straw at harvest had a SiO_2 content of 13% in Yamagata prefecture and the soil contained 13.0 mg/100 g available SiO_2 (determined by acetate buffer method). The P value was 0.8 on the soil where rice straw had a SiO_2 content of 11% and the soil contained 10.5 mg/100 g available SiO_2. By contrast, a significant regression was not obtained for the data obtained in Gifu prefecture (Figure 4.2). In Gifu, nickel slag containing a large amount of magnesium has been widely applied. Soils in Gifu are deficient in magnesium and that fertilizers containing magnesium are effective in these soils. The effect of magnesium in the nickel slag is considered to have masked the effect of Si in the slag.

The typical P-T curve obtained from the data for Yamagata prefecture, shows that a profitable increase in rice yield was observed when the rice straw had a SiO_2 content of less than 11%, while no effect was observed when it had a SiO_2 content higher than 13%. The SiO_2 content between 11-13% was the critical line for the necessity of silicate application.

From these data and additional data from other prefectures, the criterion for application of silicate fertilizers was established (Table 4.1). This criterion is divided into three classes. A profitable increase in rice yield could be expected with the highest probability if rice straw had a SiO_2 content of less than 11%, or the amount of soluble SiO_2 in soil is lower than 10.5 mg/100 g soil (class I). If rice straw had a SiO_2 content between 11 and 13%, or the amount of available SiO_2 in soil was between 10.5 and 13 mg/100 g soil, the necessity of application of silicate fertilizer depends on the weather, occurrence of diseases and pests and other factors (class II). No increase in rice yield could be expected when the rice straw had a SiO_2 content higher than 13% or available SiO_2 content higher than 13 mg/100 g soil (class III).

These criteria vary with the year and location. For example, in the year with unseasonable weather and high occurrence of diseases or pests, the criteria for class II should be used, but in a year with nice weather, the criteria for class I should be used. In the area where other elements such as magnesium in addition to Si are also effective, these criteria are difficult to be adopted.

From data of several sources, the average SiO_2 content in rice straw for the whole country was estimated to be about 11% in 1958. This means that about 50% of paddy soils (3 million hectares) in Japan requires application of silicate fertilizers.

Table 4.1
Criteria for predicting slag (Ca-silicate) need for paddy fields

Class	SiO_2 content in rice straw at harvest	Content of available SiO_2 in soil	Response to slag application
I	11% >	10.5 mg>	A profitable increase in rice yield is expected with the highest probability.
II	11~13 %	10.5~13 mg	A profitable increase in rice yield is expected in many cases, but not in some cases.
III	>13 %	>13 mg	No profitable increase in rice yield is expected.

These criteria have been used as a guide for the application of silicate fertilizers for a long time. However, with successive applications of silicate fertilizers and other changes in cultivation environment, the method for determination of the content of soluble Si in the soil is required to be modified as described previously. The amelioration target for the amount of Si in the paddy soils (acetate-buffer method) was revised to higher than 15 mg SiO_2/100 g soil in 1987. A criterion will be established using new determination method (incubation under submerged condition and/or phosphate buffer method) in near future.

The minimum Si content of rice straw for avoidance of ear blast was found to be 11% SiO_2 at mature stage in north-east district (Sumida, 1996). However, this value was evenly 13% in 1958. Compared to 1958, the percentage of early season cultivars increased at present. Moreover, Si uptake by rice differs with areas and cultivars, therefore, a local criterion seems necessary in the future.

4.2 FIELD EXPERIMNTS ON THE EFFECTS OF SILICATE FERTILIZER APPLICATION

4.2.1 Slag-calcium silicate

Numerous trials on application effects of calcium silicate were carried out systematically at agricultural experiment stations all over the country from 1953 to 1962 (Calcium Silicate Fertilizers Association, 1969). As a result, it was found that the application effects of Ca silicate were generally obvious in degraded paddy fields and peaty paddy fields.

The effects of application of Ca-silicate were also observed in other types of soils. One important effect was that application of silicate fertilizer raised the

Table 4.2
Effect of calcium-silicate application on the optimum level of nitrogen application for rice plant
a Trial from Niigata Agricultural Experiment Station (1956)

N kg/ha	Brown rice t/ha	
	-Ca-silicate	+ Ca-silicate
0	3.10	
47	3.90 (100)	4.68 (120)
56	4.07 (100)	4.90 (120)
75	3.70 (100)	5.19 (140)
94	-	5.19

Application rate: P_2O_5 ,56 kg/ha; K_2O, 68 kg/ha; + Ca-silicate, 1.88t/ha
soil type, degraded paddy soil

b Trial from Hokkaido Agricultural Experiment Station (1957)

N application rate (kg/ha)	-Ca-silicate		+ Ca-silicate	
	Brown rice (t/ha)	% of culm infected with blast	Brown rice (t/ha)	% of culm infected with blast
0	5.10 (100)	3.7	5.19 (102)	0.8
37.5	5.13 (100)	18.4	5.44 (106)	7.1
46.9	4.73 (100)	42.2	5.33 (113)	4.8
56.3	5.00 (100)	37.6	5.69 (114)	26.1
65.6	4.67 (100)	43.5	5.69 (122)	16.0

Application rate: P_2O_5, 56.2kg/ha; K_2O, 48.7 kg/ha; + Ca-silicate, 1.69t/ha
soil type, peat soil

optimum level of nitrogen fertilizers (Table 4.2a, b). The increased yield of rice in Japan after World War II has been ascribed to the increased application of nitrogen fertilizers, and the introduction of new cultivars with short culms and panicle number type, together with silicate application made heavy application of nitrogen fertilizers possible. Excess application of nitrogen fertilizers causes lodging, increases the disease susceptibility of rice, and causes mutual shading of leaf blades leading to a decrease in yield. The application of Si, however, protects against these stresses. That is why Si application increases the optimum application rate of nitrogen fertilizers.

The application rate of silicate fertilizers is determined based on the amount of Si taken up by rice. Usually 1.5 to 2.0 tons of calcium silicate per hectare is applied in many areas. At the end of the 1960's, the annual consumption of

Table 4.3
Effect of continuous application of calcium silicate on the rice yield (brown rice t/ha)

a Tokushima Agricultural Experiment Station

Year	Calcium silicate (t/ha)		
	0 (A)	1.68 (B)	B/Ax100
1956	3.53	3.57	101
1957	3.72	4.09	110
1958	3.82	4.13	108
1959	4.29	4.38	102
1960	4.32	4.67	108
1961	3.69	3.87	105
1962	4.04	4.73	117
1963	3.23	3.88	120
1964	3.42	3.83	112
1965	3.62	3.87	107
1966	3.18	3.78	119
1967	4.95	5.59	113
Average of 12 years	3.82	4.20	110

b Fukui Agricultural Experiment Station

Year	Calcium silicate (t/ha)			
	0	1.13	2.27	3.40
1954	4.16 (100)	4.60 (111)	4.70 (113)	4.62 (112)
1955	5.16 (100)	5.95 (115)	6.55 (127)	7.22 (140)
1956	5.54 (100)	5.77 (104)	5.58 (101)	5.62 (102)
1957	5.03 (100)	5.46 (108)	5.40 (107)	5.51 (110)
1958	5.08 (100)	5.53 (109)	5.73 (113)	5.60 (110)
1959	5.24 (100)	5.42 (103)	5.46 (104)	5.65 (108)
1960	5.15 (100)	5.50 (107)	5.47 (106)	5.73 (111)
1961	5.54 (100)	5.93 (107)	5.96 (108)	5.86 (106)
1962	5.36 (100)	5.89 (110)	5.90 (110)	5.82 (109)
1963	4.93 (100)	5.36 (109)	5.73 (116)	5.90 (120)
1964	5.30 (100)	5.65 (107)	5.71 (108)	5.95 (112)
Average of 11 years	5.14 (100)	5.55 (108)	5.65 (110)	5.77 (112)

Soil type: fine textured mottled gley soil TN, 0.28%; TC, 2.1%; N, 120 kg/ha; P_2O_5, 60kg/ha; K_2O, 120kg/ha
Available SiO_2: 10-15mg/100g; irrigation water: 14.5ppm SiO_2;
water requirement in depth: 15mm/day

c Hiroshima Agricultural Experiment Station

Year	Calcium silicate (t/ha)		
	0 (A)	1.40 (B)	B/Ax100
1955	4.60	4.82	105
1956	4.17	4.36	105
1957	4.34	4.51	104
1958	4.64	4.72	102
1959	4.56	4.56	100
1960	4.48	4.58	102
1961	4.21	4.26	101
1962	5.07	4.99	98
1963	5.74	5.77	101
Average of 9 years	4.65	4.73	102

Soil type: medium textured gray-brown soil with manganese accumulation
N, 83 kg/ha (1955-1959);120 kg/ha (1960-1963); P_2O_5 53 kg/ha;. K_2O, 75 kg/ha
available SiO_2:16mg/100g, irrigation water: 27.5ppm SiO_2 water requirement in depth: 13mm/day

calcium silicate was over one million tons. If these fertilizers were applied to degraded paddy fields (estimated to 690 thousand hectares), the application rate was 1.5 tons per hectare.

Successive application of calcium silicate affects both rice yield and soil properties, but was continued for more than 10 years at the Prefectural Agricultural Experiment Stations in Hiroshima, Tokushima, and Fukui. The average increase of rice yield was 10% in Tokushima (medium textured gray soil), and 12% in Fukui (fine textured mottled strong-gley soil) (Table 4.3a, b). In every year, the rice yield without application of calcium silicate was inferior to that with application. By contrast, in Hiroshima (medium textured gray-brown soil with Mn accumulation) (Table 4.3c), the rice yield started to decrease from the 8th year of application. This might be caused by a reduction of soil nitrogen fertility. Application of calcium silicate increses soil pH, and stimulates mineralization of soil organic nitrogen. Therefore, successive application of calcium silicate lowers the content of soil humus and total nitrogen. In addition, application of calcium silicate usually increases the exchangeable base, especially calcium.

As successive application of calcium silicate not only increases Si supply but also the amount of soil nitrogen available for rice by degrading soil organic nitrogen, the successive application causes neither the reduction of soil nitrogen nor unbalance of micronutrients in the fine textured soils rich in organic matter. However, in coarse textured soils poor in organic matter, successive application of calcium silicate will cause reduction of soil nitrogen fertility and unbalance of micronutrients, resulting in yield decrease. Therefore, organic matter needs to be applied to these soils.

The effect of calcium silicate application on the yield of sugar cane was also examined at Agricultural Experiment Stations in Kagoshima and Okinawa. Silicon was suggested to be a limiting factor of sugar cane yield in these areas, and 5% SiO_2 in top leaves was proposed as a criterion for diagnosis of Si deficiency (Furue and Nagata, 2000).

4.2.2. Porous hydrate calcium silicate

Solubility test under flooded condition, pot and field trials have shown that porous calcium silicate is more effective than commercial silicate slags in use efficiency of Si by rice (Saigusa et al., 1998a, b). Figure 4.3 shows the concentration of Si in a non-allophanic Andosol incubated under submerged conditions. The Si concentration in the soil solution was increased by addition of porous hydrate calcium silicate (PS) 2-3 times, while the addition of various commercial silicate slags (SS) increased the Si concentration in the soil solution only by 1.0-1.6 times.

In pot experiments, the shoot at the 9th leaf-stage had a Si content of about 7% when porous hydrate calcium silicate was applied at 1.50 g per kg soil (Figure 4.4), which was the highest among all treatments. The Si concentration in the soil solution sampled at the same time, was also higher in the soil supplied porous hydrate calcium silicate than other fertilizers (Figure 4.5).

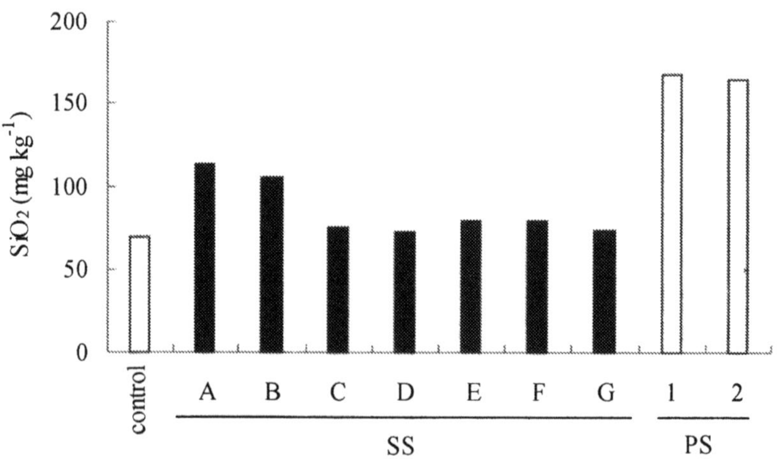

Figure 4.3. Soluble Si concentration (incubation method) in soil supplied with various silicate slags (SS) and porous hydrate calcium silicate fertilizers (PS) under submerged conditions. Amount of materials supplied was 1.50 g/kg soil.

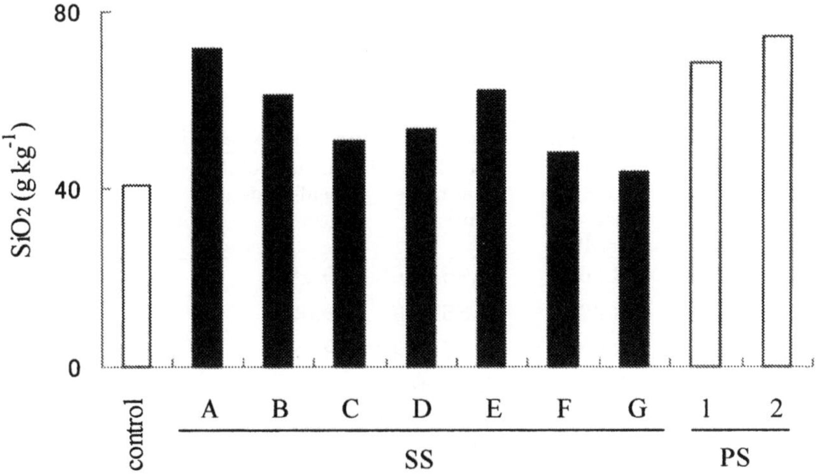

Figure 4.4. Si concentration of the rice shoots at the 9th leaf stage supplied with various silicate slags (SS) and porous hydrate calcium silicate fertilizers (PS).

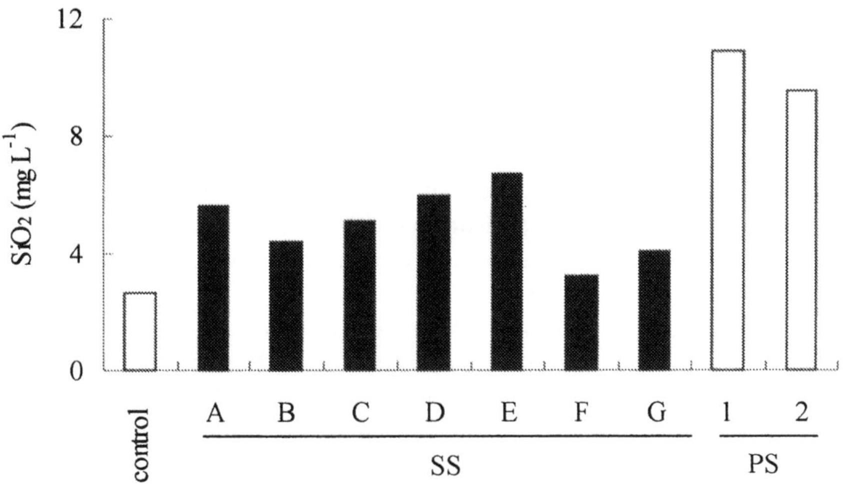

Figure 4.5. Si concentration in soil solution at the 9th leaf stage of rice plant supplied with various silicate slags (SS) and porous hydrate calcium silicate fertilizers (PS).

4.2.3. Silica gel and potassium silicate

Si in silica gel has a low solubility in water. However, as the speed of Si dissolution is fast, Si could be readily supplied to rice when the Si concentration in the solution is decreased by the uptake of rice.

Table 4.4
Effect of silica gel application on the character of rice seedlings

Amount applied (g)	Seedling height (cm)	Leaf number per seedling	Erectness of leaf-blade* (cm)	Length of 3rd leaf-blade (cm)	Dry weight of 100 seedlings	SiO_2 content of seedlings (%)
0	15.8	3.16	4.95±0.68	9.18	1.79	2.46
50	15.8	3.22	3.86±0.74	9.05	2.00	4.47
100	16.2	3.22	3.57±0.42	9.09	1.94	4.77
250	16.0	3.12	2.96±0.52	9.34	2.22	6.89
500	16.4	3.22	2.52±0.78	9.83	2.31	6.53

*expressed by distance between X and Y

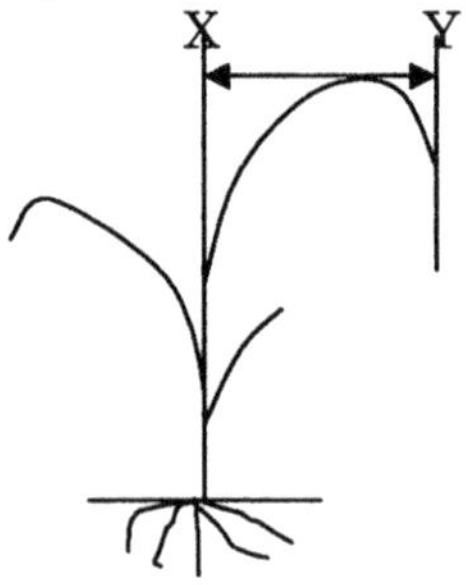

Fujii (1999) investigated the effect of silica gel application on the quality of rice seedlings and found that application of 250 g silica gel to each nursery bed (2.5 kg soil) was effective for increasing Si concentration, dry weight and erect degree of leaf blades (Table 4.4).

Silicon application to the field is known to alleviate rice blast damage. Recently, Maekawa et al. (2001) investigated the effect of the application of silica gel and potassium silicate on the occurrence of blast disease at the seedling stage. They found that application of silica gel at 200 or 250 g per nursery bed (2.5 kg soil) significantly inhibited the occurrence of blast, being 10% of the control (without application) (Figure 4.6). The application of potassium silicate at 12 g per nursery bed was also effective to suppress blast occurrence. On the other hand, the effect of fungicide (Tricyclazole) differed with application time. When it was sprayed before infection with blast, the suppressive effect was obvious, but the effect was hardly observed when Tricyclazole was sprayed after the infection.

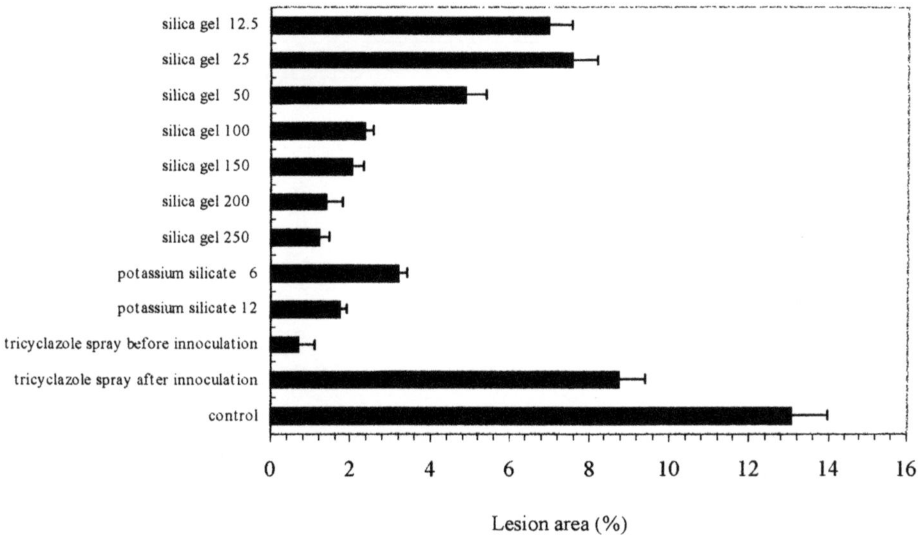

Figure 4.6. Effect of application of silica gel and potassium silicate on blast disease.

These results indicate that application of silica gel and potassium silicate to nursery bed is as effective as fungicide in controlling blast disease of rice.

4.3. EFFECT OF CALCIUM IN SLAGS ON SILICON UPTAKE BY RICE

The main composition of slags, which are applied widely, is calcium silicate. It contains not only Si, but also a large amount of Ca. Tsuno and Higashi (1984a, b), Tsuno and Kasahara (1984) reported that successive application of silicate fertilizer caused blast disease in rice plants, based on their field trials. They indicated that this resulted from an increased Ca content in leaves because they found that the leaves had a Ca content negatively correlated with their Si content, and that when the Ca content was increased without changing Si content, the number of silica bodies decreased. From these observations, they concluded that the large amount of Ca in the slag reduced the formation of silica bodies and resulted in an antagonism between Si and Ca.

However, Miyake and Takahashi (1992) failed to find a suppressive effect of Ca on Si uptake in a field experiment with various levels of Ca and Si. Table 4.5 shows a part of their results. When potassium silicate was applied at 15.5 and 62 kg SiO_2/10a, application of calcium chloride at 22.5 and 90 kg CaO/10a, respectively, affected neither the rice yield, nor Si contents in the leaf and stem,

Table 4.5
Effect of Ca application on yield and Si content of rice

	Treatment (kg SiO_2 or CaO/10a)			
K_2SiO_3	15.5	15.5	62	62
$CaCl_2$	0	22.5	0	90
Yield (kg/10a)	494	539	519	544
SiO_2 content (%)				
Leaf and stem	13.8	13.2	12.4	11.8
Hull	18.2	17.6	16.9	17.9
Ca content (%)				
Leaf and stem	0.31	0.33	0.25	0.32
Hull	0.17	0.21	0.21	0.20

and in the hull. The number of silica bodies in the flag leaf was not affected by Ca either. Combined application of calcium silicate and calcium chloride also showed a similar effect.

An experiment with solution culture was also conducted to examine the interaction between Si and Ca in terms of uptake, silica body formation, and chemical form of Si (Table 4.6). In either 0.33 or 1.66 mM Si solution, simultaneous application of 0.37 - 2.5 mM $CaCl_3$ did not affect the Si content of rice shoot (Table 4.6, Ma and Takahashi, 1993).

The uptake of Si was also not affected (Table 4.6). However, both the Ca content in the shoot and Ca uptake were decreased as the Si concentration was increased (Table 4.6). This Si-induced decrease of Ca uptake may be attributed to decreased transpiration caused by Si accumulation because Ca uptake is

Table 4.6
The percent content of Ca and Si in the shoot, the uptake amount, and the number of Si bodies in the third leaf blade of rice plants grown in a nutrient solution containing Ca and Si at various concentrations.

K_2SiO_3 (mM)	0.33			1.66		
$CaCl_2$ (mM)	0.37	1.25	2.50	0.37	1.25	2.50
	Content (%)					
Ca	0.20	0.38	0.47	0.16	0.28	0.36
Si	1.49	1.44	1.57	5.30	5.34	5.31
	Uptake (mg/pot)					
Ca	20.9	38.4	47.0	15.4	31.7	39.7
Si	153.4	144.9	157.2	526.1	596.6	589.0
	Si bodies (no./1 mm^2)					
	4.7	4.6	2.5	61.1	69.2	62.3

closely related to the transpiration. In addition, Si is also deposited in the free space of the roots (Sangster, 1978). This may also affect Ca uptake by decreasing apoplastic flow.

The chemical forms of Si in rice leaves were also unaffected by the increase in Ca level (Figure 4.7). Silica gel was the most prevalent form at any Ca level, and the monomeric and colloidal Si remained below 8.0 mM Si regardless of Ca and Si levels. The number of silica bodies detected by soft x-rays in the third leaf blade was unaffected (Table 4.6).

All these results indicate that Ca affects neither the Si uptake, Si form, nor the formation of silica bodies.

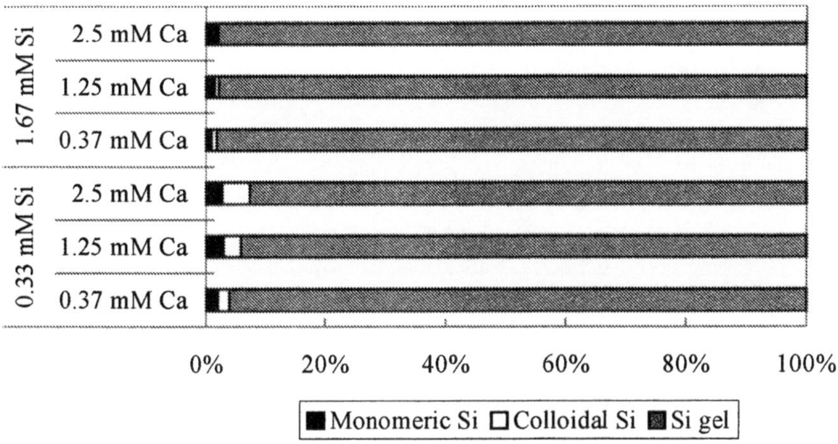

Figure 4.7. Percentage of three Si forms in the leaf blade of rice plants supplied Ca ($CaCl_2$) and Si (H_4SiO_4) at various levels.

Chapter 5

Silicon-accumulating plants in the plant kingdom

5.1. CRITERIA FOR DISCRIMINATING SI-ACCUMULATING PLANTS FROM NON-ACCUMULATING PLANTS

The mineral composition of plants varies with the plant species and the growth environment. Striegel (1912) compared the mineral compositions of various plant species growing on the same soil and found the largest variation in the contents of Si and Ca. In the monocot species (all gramineous plants) tested, the Si content was high and the Ca content low. In dicot species (including various families), however, the Ca content was high and Si content low. There was a clear difference in a ratio of Ca to Si between monocots and dicots.

In an attempt to characterize Si-accumulating plant species in the plant kingdom, Takahashi et al. (1976-1981) made an extensive survey on the mineral composition of nearly 500 plant species ranging from *Bryophyta* to *Angiospermae*. Table 5.1 shows the water soluble Si contents in the soils from which samples were collected.

Si-accumulating plants can be discriminated from non-accumulating plants, using two criteria (Si content and Si/Ca ratio, Table 5.2). The Si content of 0.5% is used as the critical value. This value is based on the following supposition. If the Si concentration in the soil solution (1-12 ppm) is 10 ppm

Table 5.1
Content of water soluble Si in soils where plant samples were collected

Sampling sites	Water soluble SiO_2 (mg/100g dry soil)
Nippon Shinyaku Botanical Gardens*	3.9
Kyoto Prefectural Botanical Gardens*	2.1
Okayama University Experimental Farm*	2.7
Hiroshima Pref. Agric. Experiment Station*	2.2
National Institute of Genetics Farm**	7.9

* alluvial soil ** volcanic ash soil

Table 5.2
Criteria for Si-accumulating plants

	Type		
	Si-accumulator	Intermediate	Si-excluder
Si content (%)	>1.0	1-0.5	<0.5
Si/Ca ratio	>1.0	1-0.5	<0.5
Degree of Si accumulation	+	±	-

and water requirement (300-700 ml per 1 g dry matter production) is 500 ml, the content of Si taken up by passive absorption on a dry weight basis will be 0.5%. Since the plants that accumulate Si tend to have a low Ca concentration, the Si/Ca ratio is used as the second criterion.

Plants with a Si content and Si/Ca ratio higher than 1.0% and 1.0, respectively, are defined as Si accumulators (Table 5.2). By contrast, plants with a Si content lower than 0.5% and Si/Ca ratio lower than 0.5, are defined as Si excluders (non-accumulating plants). Plants with a Si content and Si/Ca ratio in between Si accumulators and excluders are intermediate type plants. For example, if a plant has a Si content higher than 1.0%, but the Si/Ca ratio is below 1.0, this plant belongs to the intermediate type.

5.2. CHARCTERISTICS OF SILICON ACCUMULATORS AND THEIR DISTRIBUTION IN PLANT KINDOM

From an analysis of the mineral composition, 175 species collected from Nippon Shinyaku Botanical Gardens were classified into three groups based on their Si content (Table 5.3). In group A (shown by + in Table 5.3), the Si content was more than 1.5%. This group includes monocots I, *Pteridophyte* I, and *Bryophyte*. In group B (shown by – in Table 5.3), the Si content was lower than 0.25%. This group includes monocots II, dicots II, *Gymnosperm*, and *Pteridophyte* II. In group C (shown as ± in Table 5.3), the Si content and Si/Ca ratio was 0.86% and 0.54, and dicots I were included (Table 5.3). These results suggest that the plants in group A take up Si actively (active uptake type), but the plants in group B reject the uptake of Si (rejective uptake type). Plants in group C may take up Si passively (passive uptake type).

According to the criteria shown in Table 5.2, some monocots, some *Pteridophytes* and *Bryophytes* tested are Si accumulators (Table 5.3). Among the monocots, most Si accumulators belong to *Cyperacea* and *Gramineae* (for details, see Appendix-3A).

Table 5.3
Mineral composition of 175 plant species collected from Nippon Shinyaku Botanical Gardens

Classification of plant species		Number of species	Si (%)	Ca (%)	Mg (%)	K (%)	B (ppm)	Si/Ca	
Angiospermae		147	0.50	1.66	0.24	2.70	19.2	0.61	
Monocots	I	22	1.88	0.65	0.14	2.39	3.0	3.12	+
	II	40	0.20	1.78	0.25	3.18	13.7	0.14	-
Dicots	I	8	0.86	1.76	0.17	2.34	9.5	0.54	±
	II	77	0.23	1.87	0.27	2.57	27.7	0.15	-
Gymnospermae		12	0.13	1.20	0.11	1.15	26.7	0.17	-
Pteridophyta	I	10	1.88	1.13	0.26	2.17	8.5	3.68	+
	II	4	0.20	0.80	0.27	1.64	28.9	0.23	-
Bryophyta		2	3.46	1.06	0.26	0.93	7.6	3.31	+
Total		175	0.58	1.57	0.23	2.52	18.9	0.76	

Table 5.4 shows the contents of Si and Ca in 45 species of *Pteridophyta* collected from Kyoto Prefectural Botanical Gardens. All species in *Lycopsida* and *Equisetopsida* were Si accumulators, while both Si accumulators and non-accumulators were included in *Filicales*, the largest order of *Filicopsida* (Table 5.4, for details, refer to Appendix-3B). There were no families that had both Si accumulators and non-accumulators (see Appendix-3B).

The analysis of 10 plant species in *Cyperaceae* collected from Nippon Shinyaku Botanical Gardens revealed that *Cyperus* and *Carex* are Si accumulators while *Scirpus* is a Si non-accumulator (Table 5.5, for details see Appendix-3A). Table 5.6 shows the contents of Si and Ca in 211 gramineous species collected from Kyoto Prefectural Botanical Gardens and National Institute of Genetics Farm. All *Gramineae* species are Si accumulators, but the degree of Si accumulation differs among subfamilies. The degree of Si accumulation was in the order of *Bambusoideae>Pooideae>Panicoideae> Eragrostoideae* (Table 5.6, for details, see Appendix-3C, D).

Commelinaceae and *Juncaceae*, which are close to *Gramineae* and *Cyperaceae* were also analyzed for their Si and Ca contents. These samples were collected from Kyoto Prefectural Botanical Gardens and Hiroshima Prefectural Agricultural Experiment Stations. The results showed that *Commelinaceae* had a relatively high Ca content, but some species had a higher Si content than Ca content (Table 5.7, for details, refer to Appendix-3E, F), suggesting that Si accumulators also exist in addition to Si excluders and intermediate type in this family. Species of *Juncaceae* contain small amounts of Si and Ca

Table 5.4
Contents of Si and Ca in 45 species of *Pteridophyta* collected from Kyoto Prefectural Botanical Gardens

	No. of Species	Si %	Ca %	Si/Ca
Si accumulator	23	3.01	1.38	2.68
Lycopsida	2	4.60	0.55	8.42
Equisetopsida	2	5.81	2.18	2.79
Filicopsida				
Marattiales				
Marattiaceae	1	1.66	1.23	1.35
Filicales				
Osmundaceae	2	4.01	1.04	4.17
Blechnaceae	1	2.35	1.11	2.12
Pteridaceae	4	2.67	0.84	3.05
Thelypteridaceae	4	3.15	1.98	1.70
Athyriaceae	7	1.86	1.52	1.20
Si non-accumulator	22	0.26	1.26	0.21
Filicopsida				
Filicales				
Dryopteridaceae	17	0.27	1.42	0.18
Davalliaceae	2	0.37	1.24	0.44
Polypodiaceae	3	0.12	0.80	0.15
Total	45	1.66	1.33	1.25

and some of them are ranked as excluder but on average *Juncaceae* is intermediate type.

Table 5.5
Contents of Si and Ca in *Cyperaseae* collected from Nippon Shinyaku Botanical Gardens

	No. of Species	Si %	Ca %	Si/Ca
Cyperaseae	10	1.62	0.60	3.44
Si accumulator				
Cyperus	3	2.06	0.64	3.37
Carex	5	1.91	0.60	4.85
Si excluder				
Scirpus	2	0.21	0.52	0.41

Table 5.6
Contents of Si and Ca in *Gramineae*

	No. of species	Si %	Ca %	Si/Ca
Bambusoideae				
*Oryzeae**	73	7.36	0.44	17.9
Others**	78	3.91	0.69	5.8
*Pooideae***	22	2.69	0.63	4.6
*Panicoideae***	30	2.73	0.90	3.4
*Eragrostoideae***	8	1.73	0.84	2.5
Total	211	4.73	0.63	9.4

* sampled from National Institute of Genetics
** sampled from Kyoto Prefectural Botanical Gardens

There are no species belonging to Si accumulators in dicots collected (Table 5.3), but the Si content has been reported to be high in *Urticaceae*. In addition, it was observed that cucumber cultured hydroponically with high Si concentration gave a high Si content. To characterize the degree of Si accumulation in these families, samples of *Urticaceae* and *Cucurbitaceae* were collected from Kyoto Prefectural Botanical Gardens and Experimental Farm of Okayama University. These plant species contained over 1% Si, especially species *Cucurbitaceae* showed a higher Si content (Table 5.8). As these species had a much higher Ca content than Si content, they belong to the intermediate type according to the criteria described above (see Appendix-3G, H for details).

Table 5.7
Contents of Si and Ca in *Commelinaceae** and *Juncaceae***

	No. of Species	Si %	Ca %	Si/Ca
Commelinaceae	28	1.23	2.32	0.56
Si accumulator	4	2.90	1.83	1.52
Intermediate	17	1.24	2.61	0.50
Si excluder	7	0.25	1.90	0.15
Juncaceae	12***	0.33	0.29	1.12

* Collected from Kyoto Prefectural Botanical Gardens
** Collected from Hiroshima Prefectural Agricultural Experiment Station
*** Number of cultivars of *Juncus*

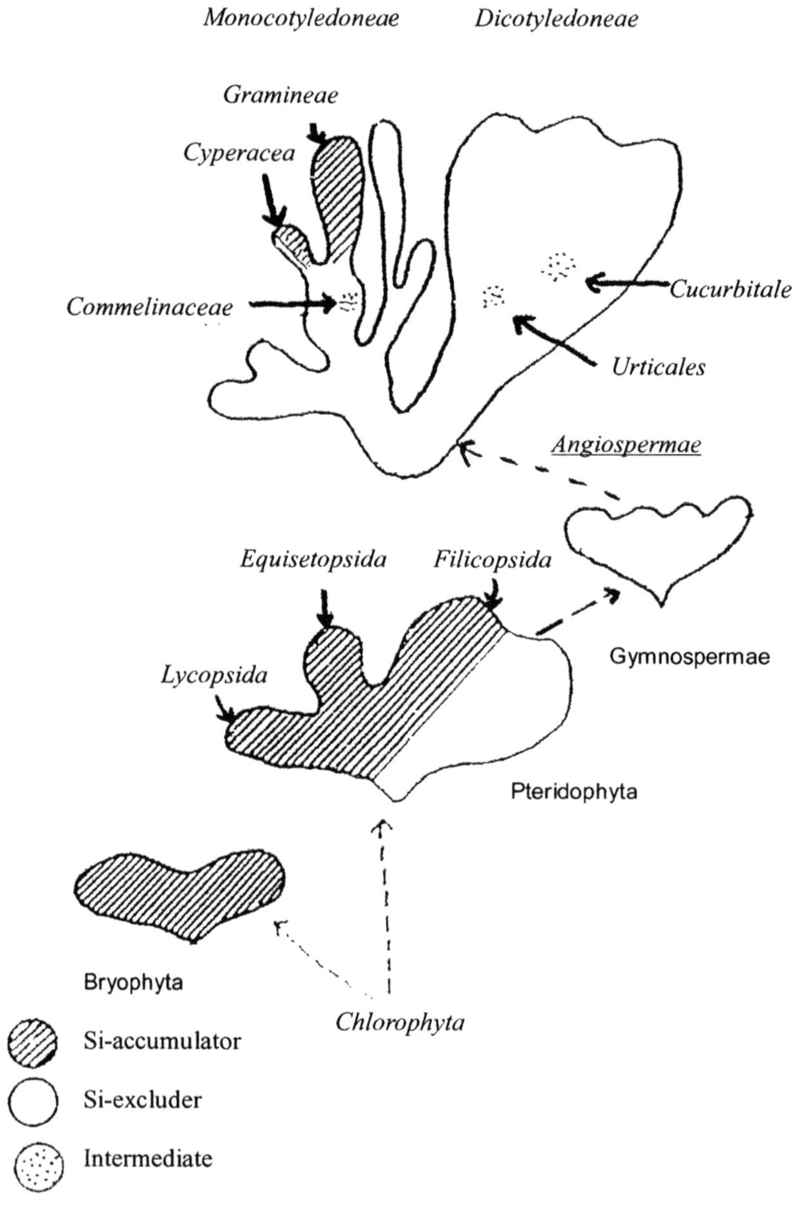

Figure 5.1. Distribution of Si-accumulators in phylogenetic tree.

Table 5.8
Ccontents of Si and Ca in *Cucurbitaceae** and *Urticaceae***

	No. of species	Si %	Ca %	Si/Ca
Cucurbitaceae	8	2.09	4.40	0.47
Urticaceae	5	1.03	4.64	0.32

* collected from Okayama University Experimental Farm
** collected from Kyoto Prefectural Botanical Gardens

Figure 5.1 shows the distribution of Si accumulators in the phylogenetic tree constructed based on these data. Silicon is highly accumulated in *Bryophyta*, and *Lycopsida* and *Equisetopsida* of *Pteridophyta*, but decreased from *Filicopsida* in *Pteridophyta* to *Gymnospermae* and *Angiospermae*. However, high Si accumulation is seen again in *Cyperaceae* and *Gramineae* in monocots (Figure 5.1).

It is well known that some plant species accumulate a large amount of Na, Al, Mn, Se, etc. However, only the distribution of Si accumulators is fitted well to the phylogenetic tree. Accumulation of Na, Al, Mn, and Se is related to soil factors such as salinity, acidity, reduction degree, and parent material, and their accumulation is the result of adaptation in these special soil environments. By contrast, Si is always abundant in soil. Therefore, Si accumulation depends on whether the plant takes up Si or not. From this point of view, Si accumulation is an advantageous trait for plants, and this trait is considered to be preserved.

5.3 VARIETY DIFFERENCE IN SILICON CONTENT IN THE SI-ACCUMULATING AND INTERMEDIATE-TYPE SPECIES

There are wide variations in Si content with the species as described above. Takahashi et al. (1981b) further investigated the variation in Si content among different varieties in the same species. The mineral content in the leaves of 38 *Oryza perennis* varieties growing in the same soil (Experimental Farm of the

Table 5.9
Mineral content of the leaves of 38 varieties from *Oryza perennis*

	Average	Range
Si (%)	7.67	10.60 - 5.38
K (%)	2.25	3.19 – 1.36
Ca (%)	0.39	0.57 – 0.26
Mg (%)	0.12	0.20 – 0.08
P (%)	0.16	0.34 – 0.06
S (%)	0.14	0.32 – 0.07

National Institute of Genetics) were analyzed. The Si content in leaves varied from 5.38 to 10.60% and was 7.67% on average (for details see Appendix-3C). The variation in Si content was rather small compared with that of K, Ca, Mg, P, S (Table 5.9). The relationship between Si content and the content of other elements except P was not clear. In the case of P, a weak negative correlation was observed.

Recently, Ma et al. (2002c) analyzed the Si content of barley grain of about 400 varieties. The hull of gramineous grain usually contains a large amount of Si as does the leaves, and can also be used for the analysis of Si accumulation. The grains of two varietal groups were used; 274 standard varieties (SV)

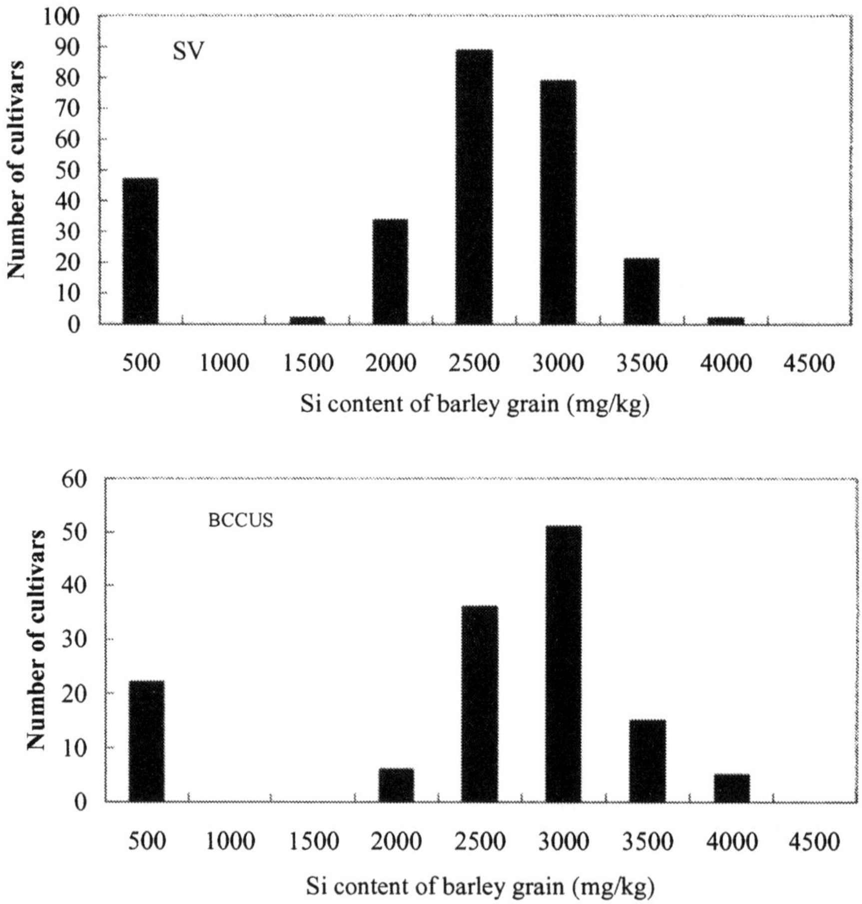

Figure 5.2 Frequency distribution of Si content of barley grains from Standard Variety (SV) and Barley Core Collection of United State (BCCUS).

Table 5.10
Difference between Si contents of covered and hull-less barley grain and of two-row and six-row barley grains

	Si content (mg/kg)	
	SV	BCCUS
Covered barley	2439.8±439.5	2640.9±468.0
Hull-less barley	26.8±23.4	116.0±44.8
Two-row	2219.3±624.8	2027.8±868.0
Six-row	1902.9±1013.2	2287.2±1078.6

selected at the Barley Germplasm Center of the Research Institute for Bioresources, Okayama University, and 135 varieties from the Barley Core Collection of United State. These barley gains were collected from the plants growing on the same soils.

The Si content of barley grain showed a large difference, ranging from 0 to 0.36% in SV and from 0 to 0.34% in BCCUS (Figure 5.2, for details, see Appendix-4A, B). The Si content was much lower in hull-less barley than in covered barley (Table 5.10). This is because most Si was localized in the hull (Table 5.11). The Si content of the hull was between 1.53 and 2.71% in the varieties tested. The Si content of two-row barley was similar to that of six-row barley (Table 5.10), suggesting that the Si content is not affected by spike row. The Si content of barley grain also did not differ with the origin of barley. Further analysis of 210 cultivars from East Asia also showed similar trends as SV and BCCUS (see appendix 4-C).

The results of the studies on rice and barley indicate that Si content also varies with the variety in the same species. However, the mechanisms responsible for the variations remain to be examined in the future. The variation might result from different capacity of uptake by the roots, and/or accumulation.

Table 5.11
Localization of Si in barley grain

SV No.	Si content (%)			Percentage of
	Total	Hull	Hulled grain	Si in the hull
39	0.25±0.03	1.99±0.23	0.10±0.01	65.9±2.7
55	0.22±0.02	2.49±0.23	0.05±0.00	78.5±0.9
110	0.31±0.01	2.71±0.45	0.06±0.00	80.6±1.6
137	0.21±0.01	1.92±0.13	0.05±0.00	77.6±1.9
211	0.23±0.01	1.99±0.20	0.03±0.00	88.1±2.0
213	0.31±0.00	2.21±0.04	0.04±0.00	89.5±2.2
223	0.18±0.00	1.53±0.07	0.03±0.00	86.4±0.6

Chapter 6

Silicon uptake and accumulation in plants

The uptake and accumulation of Si have been studied mainly using rice, a typical Si-accumulating plant. The tops of rice accumulate up to 10% Si in dry weight and many studies have indicated that rice has a specific system for transporting Si. This chapter summarizes the research on Si uptake and accumulation mainly in rice.

6.1. THREE MODES OF UPTAKE FOR SILICON

As stated in chapter 5, the Si content of the top greatly varies with the plant species, ranging from 0.1 to 10% in dry weight. The capacity of Si uptake in tomato and rice has been measured to examine the mechanism responsible for such a large variation in Si content.

Table 6.1 shows the Si concentration of rice and tomato which were grown in the same nutrient solution containing 100 ppm SiO_2 as silicic acid (Miyake and Takahashi, 1976a). The Si content in the rice tops was 20-fold of that in the roots, while the Si content in the tomato tops was one tenth of that in the roots. Although the Si concentration supplied in this experiment was higher than that in nature, the results showed that rice roots can take up large amounts of Si. The Si could be translocated from the roots to the tops in rice, while in tomato Si can not be taken up and translocated from the roots to the top. This implies that the large difference in the Si content in the tops between rice and tomato results from the difference in the ability of the roots to take up Si.

Table 6.1
Si concentration in the tops and the roost of rice and tomato plant grown in the same nutrient solution containing 100 ppm SiO_2

	Si content (%)	
	Rice	Tomato
Top	7.28	0.05
Root	0.35	0.56

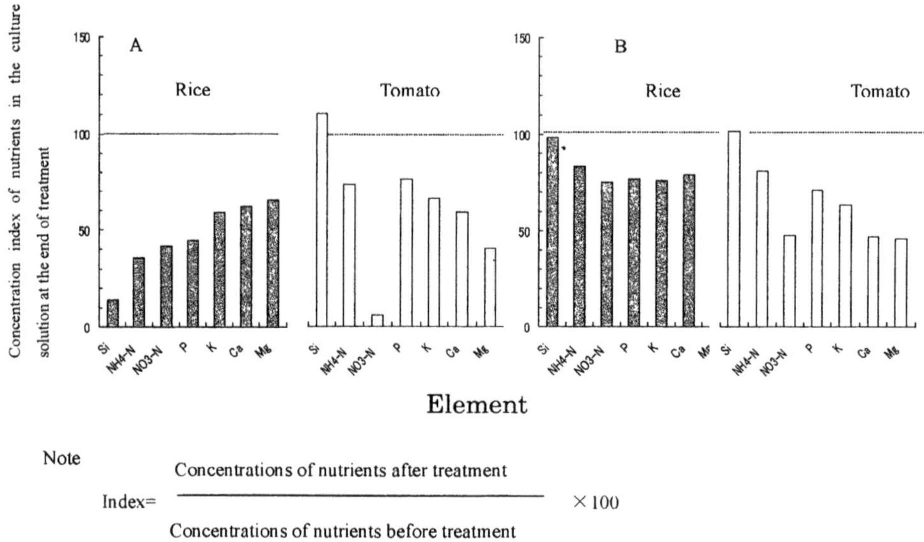

Figure 6.1. Differences in nutrient uptake between rice and tomato plants. A, intact plants; B, excised tops.

Changes in the Si concentration in the nutrient solution during plant culture supports this speculation (Figure 6.1, Okuda and Takahashi, 1962d). In this figure, the concentration index, which means (concentration of nutrients after plant culture/concentration of nutrients before culture x 100) is shown. Therefore, if the nutrients were taken up by passive transport, the concentration index should be 100. If the nutrient uptake by the roots is limited, the index should be over 100. By contrast, if the nutrients are taken up by active transport, the index should be below 100. Thus, the concentration index can reflect the nutrient-absorbing ability of the roots.

As shown in Figure 6.1A, the concentration index of Si was very low in rice, while it was over 100 in tomato. These results suggest that Si is taken up through the roots by active transport in rice, but that the uptake is limited in tomato.

Although there is a large difference in the Si-absorbing ability between rice and tomato, the difference disappeared when the roots were cut off (Figure 6.1B). This result further suggests that the difference in Si absorbing ability between rice and tomato is caused by the difference in absorbing ability of the roots.

The Si content in the bleeding sap from the cut end of stem in rice was compared with that in tomato pre-cultured in a nutrient solution containing various concentrations of silicic acid (Figure 6.2, Okuda and Takahashi, 1962d).

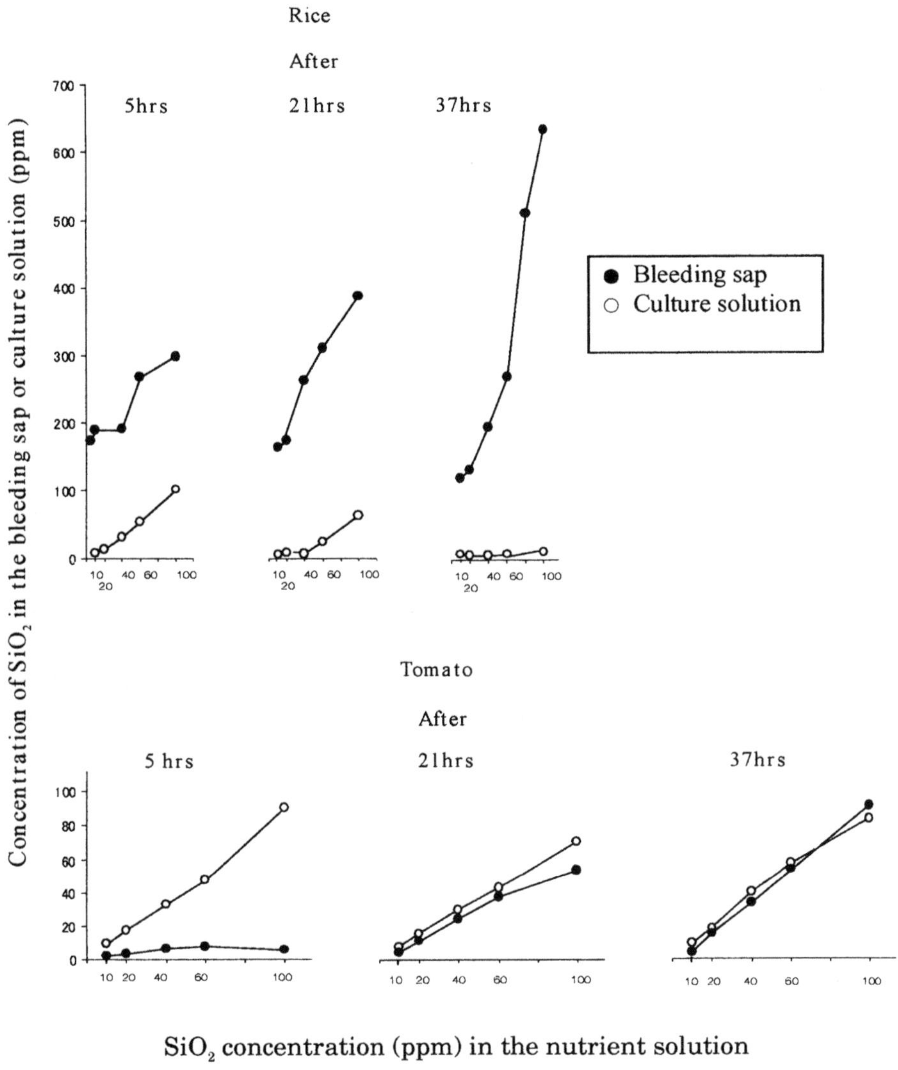

Figure 6.2. Silicon concentration in the bleeding sap from the cut ends of rice and tomato stems.

The Si concentration in the bleeding sap of rice was much higher than that in the external solution. With uptake by rice, the Si concentration in the nutrient solution decreased, resulting in several hundred-fold higher Si concentration in the bleeding sap after a 37-h culture. By contrast, the Si concentration in the bleeding sap of tomato was lower than the initial Si concentration in the external solution, and became similar to that in the external solution after a

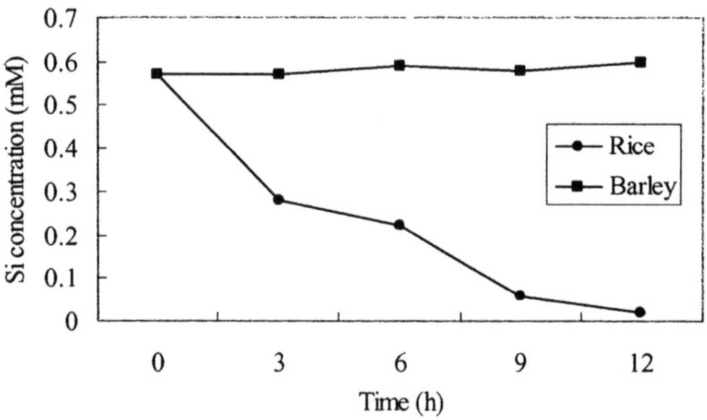

Figure 6.3. Changes in the Si concentration in a nutrient solution initially containing 0.57 mM Si with time during the culture of rice and barley.

37-h culture (Figure 6.2). These results also indicate that rice roots take up Si by active transport, while tomato roots limit the uptake of Si.

In contrast to the culture of rice which caused a rapid decrease in the Si concentration in the culture solution, that of barley did not change the Si concentration in the culture solution (Figure 6.3, Ma et al., 2002c). This suggests that Si is taken up by barley roots by passive transport.

All these results indicate that there are three different modes of Si uptake. The variation among plant species in the Si content of the top may be attributed to the difference in the mode of Si uptake.

6.2. CHARACTERISTICS OF SILICON UPTAKE BY RICE

6.2.1. High capacity for Si uptake

Table 6.2 shows the high ability of the rice roots to take up Si. The culture of rice for two days in tap water resulted in a significant decrease of Si concentration in the tap water (from original 25.6 ppm to 0.4-0.6 ppm) (Okuda and Takahashi, 1961a). This indicates that Si is removed by rice from the tap water as effectively as distillation with a glass still (Table 6.2). The water distilled with a glass still showed a higher concentration of Si than that distilled with a copper still because Si in the glass still could be solubilized. It is difficult to remove Si with an ion exchange resin because silicic acid is present as an undissociated molecule at around a neutral pH.

Table 6.2
Si concentration in water after a 48-h culture of rice seedlings and in water purified by various methods

Treatment	SiO_2 concentration (ppm)
Original water*	25.6
Demineralized water**	0.760
Re- demineralized water**	0.020
Distilled with a glass still	0.982
Redistilled with a glass still	0.672
Distilled with a copper still	0.005
Incubation of rice seedlings for 48 hours***	0.4-0.6

*Tap water; ** Ion exchange resin; ***Fresh weight (g) of rice seedlings: 40g

Figure 6.4 shows the Si uptake by various gramineous plants (Ma et al., 2002d). The uptake was carried out in a 0.5 mM $CaCl_2$ solution containing 0.6 mM Si as silicic acid for 24 h. Obviously, rice shows the highest uptake ability of Si.

6.2.2. Uptake form of Si

The chemical form of soluble Si varies depending on the pH of the solution. As shown in Figure 6.5, when the solution is below pH 8.0, Si is present as an undissociated silicic acid molecule [$(H_4SiO_4)_n$, n=2-3]. Therefore, in the usual

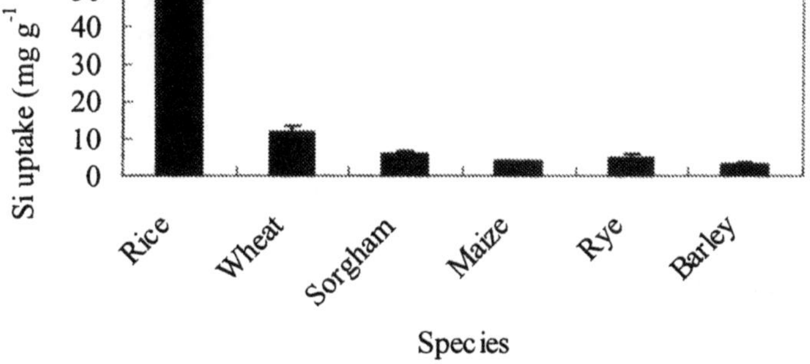

Figure 6.4. Si uptake by different gramineous plants from a solution containing 0.6 mM Si as silicic acid during a 24-h period.

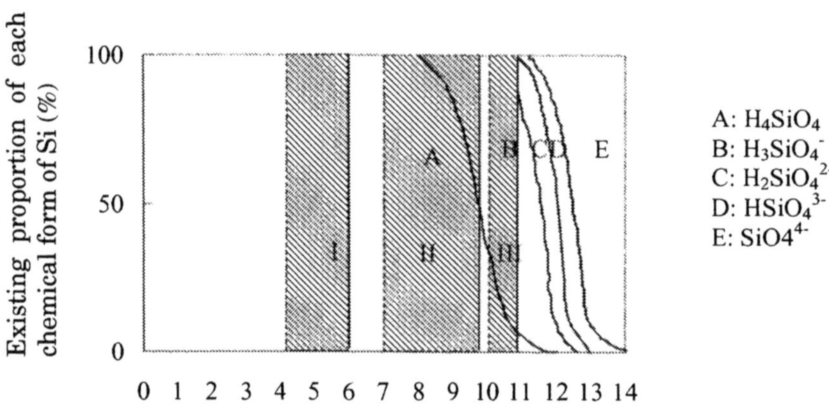

Note: Solid line: Initial pH of the solution
Broken line: Final pH of the solution
(after 24 h of the treatment)
I, II, and III indicate the range of pH during the absorption experiment, respectively.

Figure 6.5. Effect of pH of the solution on the form of dissolved Si.

culture solution and soil solution, Si is considered to be in the form of an uncharged molecule. Takahashi and Hino (1978) examined the relation between the chemical forms of Si (uncharged or ionic) and Si uptake by rice. Different forms of Si were made by changing the solution pH (Figure 6.5). Two-month-old rice seedlings were cultured in a solution containing 1 mM Si, or 1 mM Si and 1 mM P at pH 6.0, 9.7, and 11.0. At pH 6.0 (100% undissociated molecule) and pH 9.7 (50% undissociated molecule+50% ionic form), the rate of Si uptake was nearly double of that at pH 11.0 (100% ionic form) (Table 6.3). The rate of Si uptake was 4-fold that of P uptake at either pH. In addition, the P uptake was not affected by pH (Table 6.3).

The Si uptake by rice during the 24-h culture period was slightly stimulated by the presence of P at pH 6.0 and 9.7 (Table 6.3), probably due to the nutritional effect of P. However, at pH 11.0, the Si uptake was significantly inhibited in the presence of P. The uptake of phosphate ion seems to compete with that of silicate ion. These results also suggest that the uptake system for silicic acid is different from that for silicate ion in rice roots. It is concluded that rice roots take up Si in the form of non-dissociated molecule, silicic acid, and it is superior to ionic forms for uptake by rice.

Table 6.3
Effects of solution pH and coexisting P on Si uptake by rice plants*

Experimental condition			Uptake amount (mg/pot)	
			Si	P
I	pH6.0	Si only	17.7	—
		Si+P	18.7 (0.66 mmol)	5.1 (0.16 mmol)
II	pH9.7	Si only	14.7	—
		Si+P	18.2 (0.65 mmol)	5.1 (0.16 mmol)
III	pH11.0	Si only	9.2	—
		Si+P	4.5 (0.16 mmol)	4.4 (0.14 mmol)

*Cultured for 24 h. * For pH changes during the culture period: See Figure 6.5.

6.2.3. Kinetics of Si uptake

Kinetics of Si uptake was examined in rice. To investigate whether the uptake of Si is inducible, Ma et al. (2002d) cultured the rice seedlings, which were pre-cultured in a solution with or without Si in the solution containing 1.5 mM Si as silicic acid (Figure 6.6). The uptake of Si increased linearly with time, but, there was no difference in the Si uptake between the plants previously exposed to Si and not, suggesting that the uptake of Si by rice roots is not inducible.

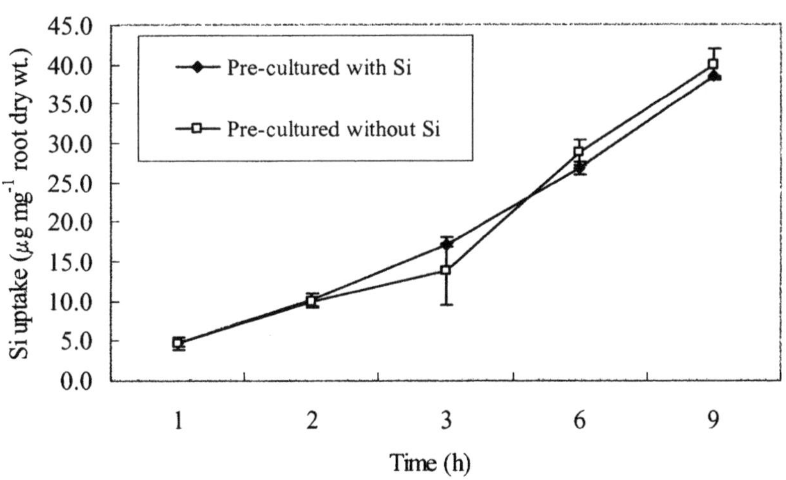

Figure 6.6. Change in the Si uptake by rice roots precultured in the solution with Si (1.5 mM Si) or without Si for 1 day. The uptake experiment was conducted in a nutrient solution containing 1.5 mM Si as silicic acid.

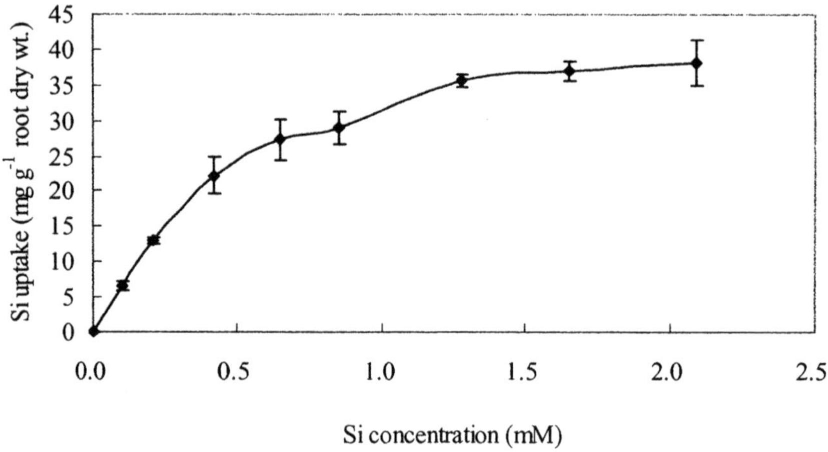

Figure 6.7. Si uptake by rice roots from a solution with various Si concentrations. The seedlings were cultured for 6 hours.

The uptake of Si by rice roots from a solution with various concentrations of Si was also investigated by Ma et al. (2002d). The uptake was saturated at 1.28 mM Si (Figure 6.7). From this uptake curve, the K_m was estimated to be 0.32 mM, suggesting that the uptake system has a low affinity for silicic acid.

6.2.4. Effect of transpiration on Si uptake

There is a "Frey-Wyssling" hypothesis for explanation of high Si content in rice. Rice originally is a water weed in tropical regions. In such regions, Si tends to be eluviated from the soil and consequently, rice absorbs a large amount of Si dissolved in water, resulting in a high Si content in the rice (Frey-Wyssling, 1930).

This hypothesis was examined by investigating the effect of transpiration on Si uptake by rice and tomato in a nutrient solution containing 100 ppm SiO_2 (Okuda and Takahashi, 1962d). When the transpiration rate was reduced by a high humidity, Si uptake by rice during 48 hours was hardly affected, but the Si uptake by tomato was decreased (Table 6.4). These results indicate that the Frey-Wyssling hypothesis is suitable for the plants which take up Si passively, but is not for the plants such as rice that take up Si actively.

However, transpiration plays a certain role in translocation and accumulation of Si to the tops of rice. The Si concentration is high in leaf blades and husks where the transpiration rate is high. Although rice roots play an important role in active uptake of Si, the Si content in the roots is much lower than that in the tops. Therefore, Si taken up by the roots is supposed to be translocated to

Table 6.4
Effect of transpiration on Si uptake by rice and tomato*

Treatment	H_2O uptake (mg/g fresh wt.)		SiO_2 uptake (mg/g fresh wt.)	
	Rice	Tomato	Rice	Tomato
Low humidity (A)	8.88	3.75	0.158	0.033
High humidity (B)	2.67	1.62	0.150	0.013
B/A	0.30	0.43	0.95	0.39

*The experiment was conducted under natural light and the uptake period was 48 hours. Fresh weight of test plants: rice A and B, 12 g; tomato A 40g, B 39g.

the shoot along with the transpiration steam and then concentrated and finally physically gelled in rapidly transpirating organs.

6.2.5. Effect of nutrient salts on Si uptake

As mentioned above, rice roots take up Si in a non-charged form of silicic acid. Therefore, the effect of nutrient ions on the Si uptake is considered to be small. However, on the other hand, it is well known that excess nitrogen causes decrease in the Si content and number of silica bodies of rice (Ishizuka and Tanaka, 1950), suggesting that Si uptake seems to be affected by some nutrient salts.

Takahashi (1982d) investigated the effect of nutrient anions and cations on the Si uptake by rice. In the presence of each nutrient at 5 mM (20 times of Si) in a nutrient solution (Kimura B solution), the Si uptake during the 72-h culture was slightly inhibited by the nitrate anion among anions (Table 6.5). However, the uptake was inhibited by 32% in the presence of ammonium-nitrogen, while other cations hardly affected the uptake of Si by rice (Table 6.5).

Table 6.5
Effects of coexisting salts on Si uptake by rice plants

Anion	Absorption index	Cation	Absorption index
Na$\underline{NO_3}$	86	$\underline{K}$Cl	96
Na$\underline{H_2PO_4}$	97	$\underline{NH_4}$Cl	68
$Na_2\underline{SO_4}$	95	$\underline{Ca}Cl_2$	97
Na$\underline{Cl}$	92	$\underline{Mg}Cl_2$	100

Note: 1. Absorption index is expressed as a ratio of Si uptake in each treatment to the control (Kimura B solution containing 0.25 mM Si).
2. Test solution: 5mM of each testing salt is added to the Kimura B culture solution supplemented with 0.25mM silicic acid. Duration of treatment: 72 h.

Table 6.6
Effects of removal of each nutrient element (for 13days) on Si uptake by rice plants

Element	Absorption index	Element	Absorption index
N	130	Mg	82
P	115	S	85
K	86	S·N	131

The effect of nutrient deficiency on the uptake of Si was also examined. When nitrogen including nitrate and ammonia nitrogen was removed from the nutrient solution for 13 days, the Si uptake by rice increased 30% (Table 6.6, Takahashi and Nishi, 1982e). Deficiency of P also caused increased uptake of Si. These results suggest that Si uptake by rice is affected by nutrient status of nitrogen and phosphorus in plant.

6.2.6. Participation of metabolism in Si uptake

Silicon uptake by rice is inhibited by H_2S and NaCN at similar extent as K and P (Mitsui et al., 1951, 1953), suggesting that Si uptake by rice is aerobic metabolism-dependent active process. The pattern of metabolism in Si uptake was investigated under various conditions by comparing with P uptake as described below

6.2.6.1. Effects of metabolic inhibitors on Si uptake

Okuda and Takahashi (1962e) investigated the effect of metabolic inhibitors on Si and P uptake (24 h) by rice in a solution containing an equimolar concentration (0.25 mM) of P and Si. The effect of metabolic inhibitors on the uptake of Si and P uptake varied with the inhibitor and can be categorized into three types (Table 6.7). The first is phlorizin (inhibitor of phosphorylation of glucose) and D-glucosamine which inhibited P uptake but not Si uptake, the second is NaF (inhibition of enolase) which inhibited both Si and P uptake similarly, and the third is 2,4-DNP (uncoupling agent of oxidative phosphorylation) and iodoacetate (inhibitor of triosephosphate dehydrogenase) which inhibited Si uptake more strongly than P uptake. Since 2,4-DNP significantly inhibited Si uptake, ATP may be required in the Si uptake by rice. It is also suggested that metabolism related to P uptake is also partially involved in the Si uptake.

The effect of metabolic inhibitors on Si uptake and P uptake was further examined in excised roots and shoots. As shown in Table 6.8 (Takahashi and Okuda, 1962e), although Si uptake by the roots was strongly inhibited by the metabolic inhibitors, the Si uptake by the excised shoots was not affected. This

Table 6.7
Effects of metabolic inhibitors on Si uptake and P uptake by rice*

	Control	NaF	2,4-DNP	D-glucosamine	Iodoacetate	Phlorizin
Si (mg)	2.79	1.41	0.48	2.69	0.38	2.89
P (mg)	1.92	1.24	0.95	1.52	1.11	1.04
H_2O (ml)	108	61	53	122	35	112
Si/P mol ratio	1.61	1.29	0.57	1.96	0.39	3.30

*One plant (panicle-forming stage) per wide neck plastic bottle (500 ml); Test solution, 3.28 mg Si/400ml (0.29mM), 3.12 mg P/400ml (0.25mM); concentration of inhibitors, 1×10^{-3} M; Incubation for 24 h, in a glasshouse late in August, fine weather.

suggests that Si uptake by the excised shoots is a passive process in which Si is translocated along with the transpiration stream. By contrast, the P uptake by both the excised roots and the excised shoots were strongly inhibited by the metabolic inhibitors, leading to leakage of P from the roots and the shoots (Table 6.8).

Mitsui and Takatoh (1961, 1962) also investigated the effect of metabolic inhibitors on Si uptake in comparison with P uptake by using radioisotope of Si (^{31}Si, the half life, 2.62 h) and P (^{32}P) for 150 minutes. The degree of inhibition of Si uptake was weaker than that of P. The degree of inhibition of P uptake was in the order of NaCN>Antimycin A, Iodoacetate, DNP>NaF>malonate> Phlorizin, while that of Si uptake followed by NaCN>DNP, Iodoacetate> Antimycin A>NaF. Malonate and Phlorizin did not inhibit Si uptake. This finding is an agreement with that observed by Takahashi and Okuda (1962e)

Table 6.8
Effects of metabolic inhibitors on Si uptake and P uptake by excised top and excised roots of rice*

Treatment	Uptake by excised top			Uptake by excised root		
	Si (μmol)	P (μmol)	H_2O (ml)	Si (μmol)	P (μmol)	H_2O (ml)
Control	0.5	1.4	3.3	6.8	3.2	1.9
NaCN	0.4	-1.5	2.9	-0.3	-6.3	2.0
2,4-DNP	0.4	-14.5	2.8	-0.2	-13.5	1.8
Iodoacetate	0.5	-13.1	3.0	-0.2	-21.9	2.0

*Three-week-old seedlings (35 seedlings in each bottle) were cut into top and root at 1cm above stem base and then set in a wide neck plastic bottle (100 ml). Each bottle contained 100 ml of 0.2 mM silicic acid and KH_2PO_4 in the presence or absence of inhibitors (5×10^{-3} M). The uptake experiment was conducted for 20 h in an incubator at 25°C under artificial light of 5000 lux.

Table 6.9
Effects of glucose, pyruvate, and acetate on the uptake of Si and P by rice during a 10-, 24- or 48-h culture

Treatment	Si (μg/pot)			P (μg/pot)		
	10 h	24 h	48 h	10 h	24 h	48 h
None	94 (100)	241 (100)	373 (100)	99 (100)	124 (100)	134 (100)
Glucose	111 (118)	247 (102)	386 (103)	151 (152)	209 (169)	249 (186)
Pyruvate	76 (81)	146 (61)	336 (90)	108 (109)	167 (135)	169 (126)
Acetate	31 (33)	155 (64)	295 (79)	108 (109)	152 (123)	187 (140)

6.2.6.2. Effect of glucose and organic acids on Si uptake

The effect of glucose, pyruvate and acetate on the Si uptake by rice was investigated by Takahashi and Okuda (1963a). Young rice seedlings which were germinated and grown in distilled water at 30°C in the dark for two weeks, were exposed to a solution containing 0.2 mM silicic acid and 0.2 mM P in the presence of 0.25 mM of glucose, pyruvate, or acetate. The uptake during a period of 48 h in the dark (30°C) showed that the uptake of P slightly increased with time in the absence of glucose, pyruvate and acetate, while that of Si significantly increased (Table 6.9). The amount of Si uptake was 3 times as large as that of P at the 48th hour. Addition of glucose significantly increased the P uptake, but not Si uptake (Table 6.9). Addition of pyruvate and acetate increased the P uptake but decreased the Si uptake although the recovery was observed at the 48th hour. These results suggest that although both Si uptake and P uptake are aerobic metabolism-dependent, the metabolic processes involved differs between Si uptake and P uptake.

Table 6.10
Effect of light on the amounts of Si and P taken up by rice during a 9-, 24- and 33-h culture

Treatment*	Si (μg) taken up during			P (μg) taken up during		
	9 h	24 h	33 h	9 h	24 h	33 h
D-D(control)	137	311	389	53	96	140
L-D	186	344	416	95	101	143
D-L	187	421	491	73	121	176
L-L	244	518	600	67	163	220
D-D+glucose	161	357	420	123	246	307

*D, darkness; L, light (5000 lux); L-D, L before exposure to Si or P and D during the exposure to Si or P. Glucose was added at 5mM.

6.2.6.3. Effects of light on Si uptake

The effect of light on Si uptake by rice was compared with the effect of glucose described above (Takahashi and Okuda, 1963b). Light stimulated the uptake of both Si and P (Table 6.10) and the longer the period of light treatment, more Si was taken up. Furthermore, the uptake was larger when the plants were irradiated during the exposure to Si or P than before the exposure to Si or P. The effect of light on P uptake was not as large as that of glucose added, while the effect of light on Si uptake was much larger than that of glucose. Light irradiation not only supplies sugars via photosynthesis but also stimulates transpiration through opening of stomata. Increased transpiration might stimulate Si uptake.

The influence of pre-exposure to light on the uptake of Si and P was further investigated. Rice was cultured in a nutrient solution containing 0, 0.1, and 0.5 mM Si or P under light (5000 lux) for 2 weeks after germination. Then the uptake of Si and P by rice from the solution containing 0.2 mM Si and P was measured under both dark and light conditions. When the plants were pre-cultured in a solution containing Si and P each at 0.1 or 0.5 mM, the rate of P uptake significantly decreased under the dark condition, while Si was slightly or not affected by light (Figure 6.8B, C). However, when the plants were pre-cultured in a solution without nutrients, the rate of P uptake did not decrease in the dark until the 5th day (Figure 6.8A), and the Si uptake was

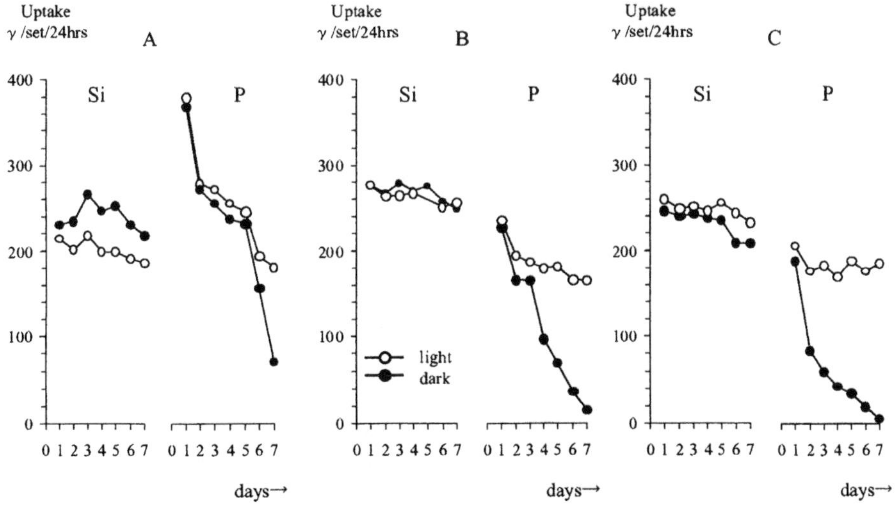

Figure 6.8. Effect of light on the uptake of Si and P by rice in a solution containing nutrient salts at 0 (A), 0.1 (B), or 0.5 mM (C) under a light intensity of 5000 lux at 25°C.

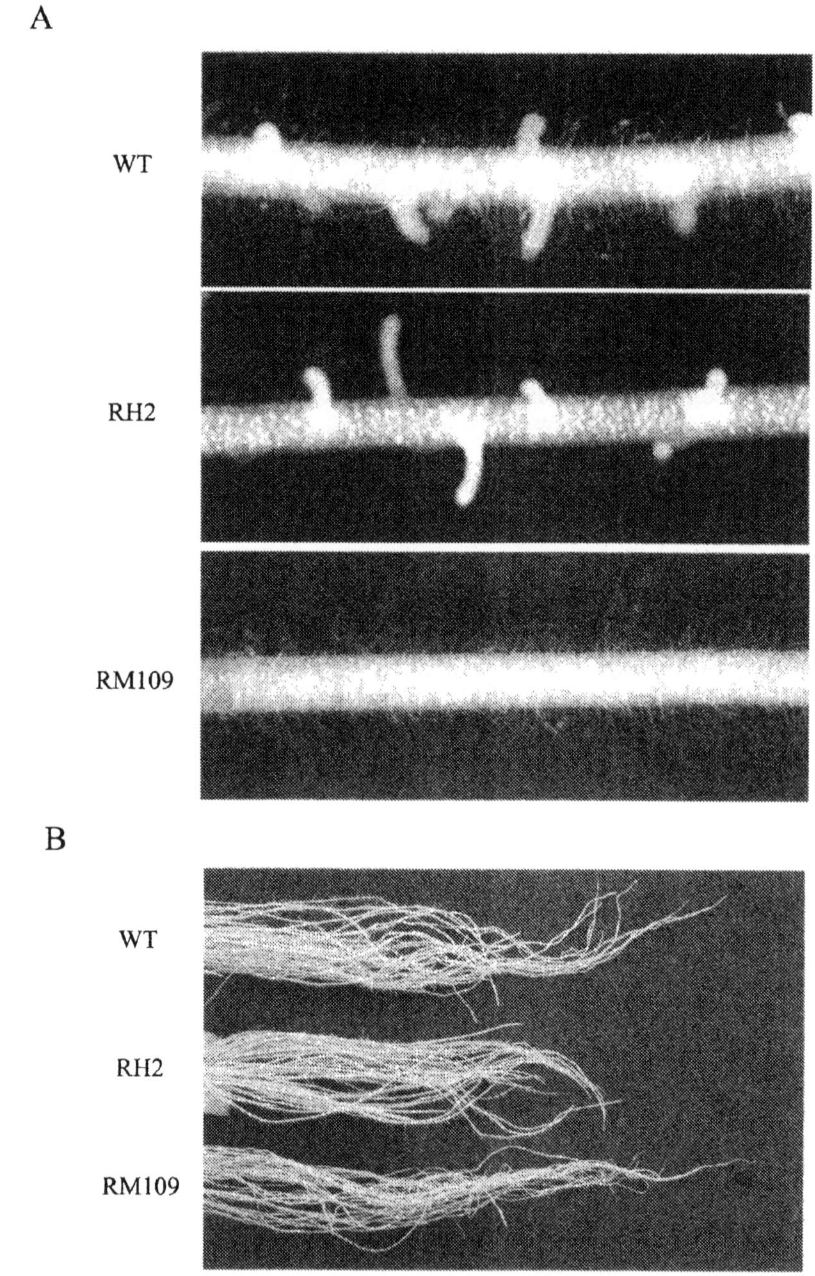

Figure 6.9. Root of WT rice cv Oochikara and two root mutants (RH2 and RM109). RH2 and RM109 are defective in the formation of root hairs and lateral roots, respectively. A, Individual root. B, Root system.

reversely reduced by light irradiation. These results suggest that P uptake is largely dependent on the supply of assimilate to the roots, while Si uptake is less dependently. The different effect of light on the uptake of Si and P also suggests that the mechanism responsible for the Si uptake in rice is different from that for the P uptake.

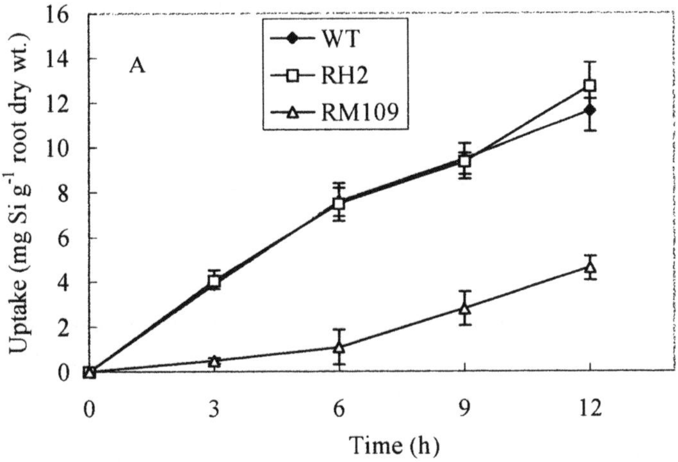

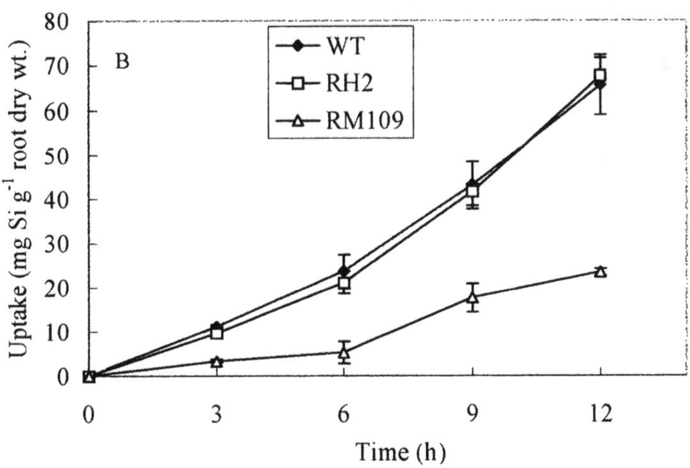

Figure 6.10. Uptake rate of Si by WT rice cv Oochikara and two mutants without root hairs (RH2) and without lateral roots (RM109). Two-week-old seedlings were placed in a nutrient solution containing 0.15 (A) and 1.5 mM (B) Si as silicic acid.

6.3. ROLE OF ROOT HAIRS AND LATERAL ROOTS IN SI UPTAKE

The root system consists of primary roots, lateral roots and root hairs. The role of root hairs and lateral roots in the Si uptake was recently investigated by using two mutants, one defective in the formation of root hairs (RH2) and the other in that of lateral roots (RM109) (Figure 6.9, Ma et al., 2001).

In a short-term experiment, the Si uptake by the wild-type rice (WT) was similar to that by RH2 in a solution with Si at either a low or high Si concentration (Figure 6.10). However, the Si uptake by RM109 was much less than that by WT. The results of long-term experiments were similar to those of the short-term experiment. The number of silica bodies formed in the third leaf in RH2 was similar to that in WT, but the number of silica bodies in RM109 was only 40% of that in WT, when grown in the soil amended with Si under flooded condition (Table 6.11). Using a multi-compartment transport box, the Si uptake at the root tip (0-1 cm, without lateral roots and root hairs) was found to be nearly the same in RM109 and WT. However, the Si uptake in the mature zone (1-4 cm from root tip) was significantly lower in RM109 than in WT, whereas no difference was found in Si uptake between WT and RH2. Root hairs have been presumed to enhance the uptake of nutrients and water by increasing the absorptive surface area However, the results indicate that root hairs do not play any demonstrable role in the Si uptake, but lateral roots largely contribute to the Si uptake in rice plant. Different from other nutrients, the Si uptake by rice roots is suggested to be mediated by a specific transport system. It seems that root hairs are defective in the specific system for silicic acid.

6.4. GENOTYPICAL DIFFERENCE IN SILICON UPTAKE

The genotypical difference in Si content has been reported to be smaller than that of other nutrients, suggesting that all cultivars of rice have a high ability to

Table 6.11
Si concentration and the number of silica bodies in a WT (cv Oochikara) of rice and two mutants, one without root hairs (RH2) and the other without lateral roots (RM109).

Treatment*	Si content shoot (mg Si g^{-1})			Number of silica bodies		
	WT	RH2	RM109	WT	RH2	RM109
-Si	20.6	21.5	17.8	19.3	19.0	12.9
+Si	32.3	34.2	19.6	71.0	82.6	26.9

*Three lines were grown in a soil amended with or without sodium silicate (2 g/kg soil) for one month. Silica bodies around 2 cm from the tip of the third leaf were counted.

take up Si. The Si content of the shoot is related to both Si-uptake ability of each individual root and development of whole root system. A comparative study on Si uptake by the individual root and root system was conducted between a *japonica* variety, Nipponbare, and an *indica* varity, Kasalath (Ma et al., 2002b). When both varieties were grown in a nutrient solution containing 0.15 mM silicic acid, the content of Si in the shoot was higher in Nipponbare than in Kasalath (Table 6.12). When grown in a solution containing 1.5 mM silicic acid, it was nearly the same in Nipponbare and Kasalath. The amount of Si taken up per plant was larger in Kasalath than in Nipponbare (Table 6.12), but the amount per g dry weight of root was higher in Nipponbare than in Kasalath. Kasalath has a larger root system than Nipponbare. These results suggest that although the Si content of the shoot is nearly the same in Nipponbare and Kasalath, different mechanisms are involved in accumulation of Si in the two varieties. The high Si content in Kasalath relies on larger root system, while that in Nipponbare on higher uptake ability per root. This speculation was supported by the results of a multi-compartment transport box experiment. Si uptake per root in Nipponbare was 30% higher than that in Kasalath (Figure 6.11).

Ma et al. (2002c) recently analyzed Si content of barley grains of about 400 cultivars grown on the same soil (refer to Appendix IV). The Si content ranged from 1240 to 3600 mg/kg in covered barley. This genotypical variation in Si

Table 6.12
Comparison of Si uptake between a *japonica* variety, Nipponbare and an *indica* varity, Kasalath*

	0.15 mM Si		1.5 mM Si	
	Nipponbare	Kasalath	Nipponbare	Kasalath
Si content (Si %)				
Shoot	1.72	1.30	4.31	4.54
Root	0.15	0.20	0.35	0.24
Dry weight (g)				
Shoot	3.36	4.10	3.08	4.86
Root	0.71	1.12	0.66	1.27
Uptake				
mg Si/plant	58.66	55.47	134.43	223.86
mg/g root dry wt.	83.59	49.82	205.40	176.89

*Two cultivars were grown in a nutrient solution containing 0.15 mM or 1.5 mM Si as silicic acid for 1 month.

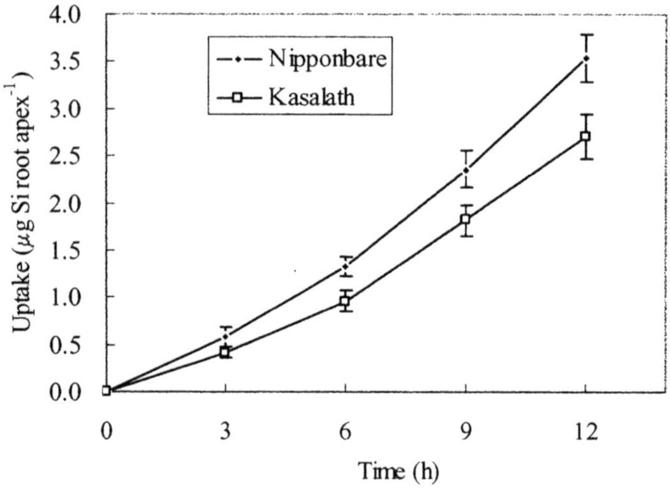

Figure 6.11. Comparison of Si uptake per root between a *japonica* variety, Nipponbare and an *indica* variety, Kasalath. Ten excised roots were placed in a compartment box and Si at 0.75 mM as silicic acid was supplied to the apical 0-3 cm of the root. The amount of Si exudated from the cutting surface was measured to determine the amount taken up.

content of barley grain may be attributed to the different ability of uptake and accumulation of Si although this remains to be examined.

6.5. A RICE MUTANT DEFECTIVE IN SILICON UPTAKE

There is an abundance of evidence as discussed above that the rice root has a transport system specific to silicic acid. However, neither the gene encoding this transporter nor that encoding the transporter protein has been isolated. A gene family encoding a Si tranporter has been cloned from marine diatom (*Cylindrotheca fusiformis*), which requires Si as an essential element (Hildebrand et al., 1993, 1997). However, similar genes were not found in rice from homology search. A rice mutant defective in active Si uptake was isolated by screening M_2 seeds (64000) of rice (cv. Oochikara) that were treated with 10^{-3} M of sodium azide for 6 h at 25°C (Ma et al., 2002a). Mutants were screened in half strength Kimura B solution containing 50 μM GeO_2. As Ge is taken up in a manner similar to Si, but is toxic to the plants, which appears as brown spots in the leaf blades, plants without brown spots in the leaves were selected. After performing progeny test for M_3 and M_4 seeds, a mutant (GR1), which showed resistance to Ge, was obtained. There were no differences in the

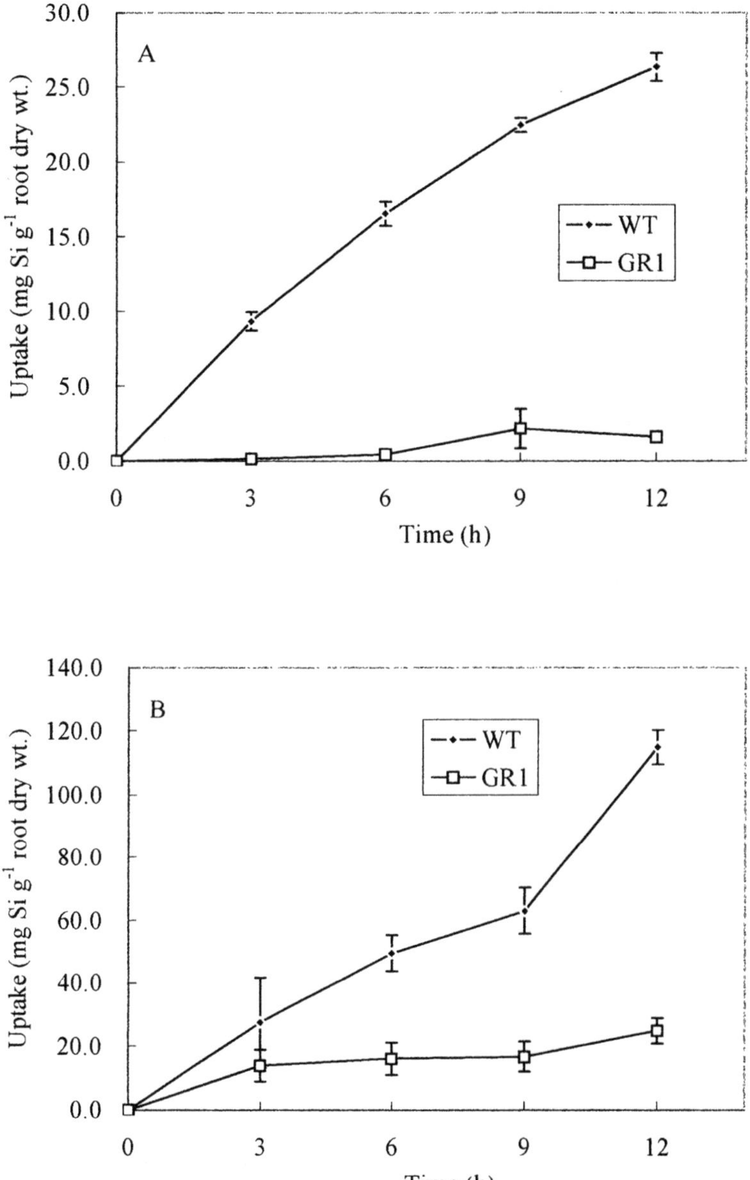

Figure 6.12. Uptake of Si by WT rice, cv Oochikara, and a mutant (GR1) defective in Si uptake. Twenty-day-old seedlings were placed in a nutrient solution containing 0.15 (A) and 1.5 mM (B) Si as silicic acid.

Table 6.13
Contents of P, K, and Si in the shoot and root of a wild type rice (cv Oochikara) and a mutant (GR1) defective in Si uptake

	Shoot		Root	
	0.15 mM Si	1.5 mM Si	0.15 mM Si	1.5 mM Si
P content (%)				
WT	0.57	0.54	0.23	0.21
GR1	0.51	0.57	0.20	0.25
K content (%)				
WT	3.08	3.18	1.02	1.14
GR1	3.12	3.23	1.01	1.17
Si content (%)				
WT	1.46	4.62	0.03	0.12
GR1	0.26	1.43	0.04	0.08

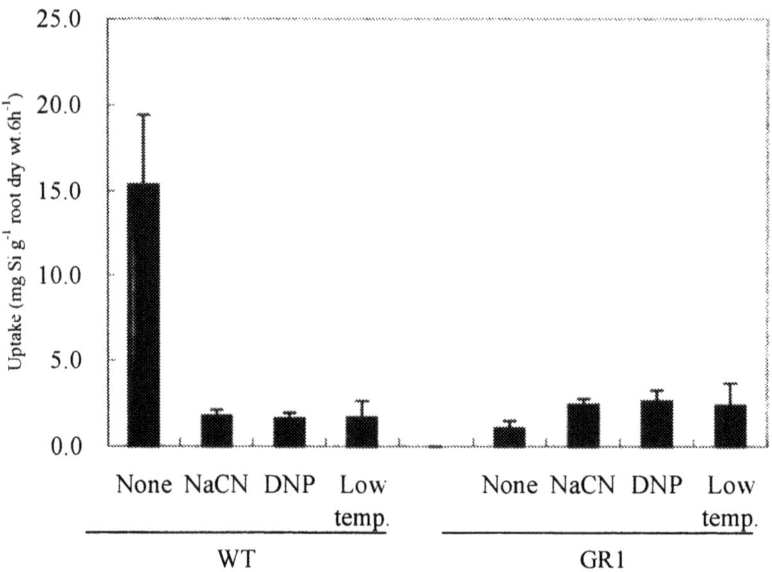

Figure 6.13. Effect of metabolic inhibitors and a low temperature on the Si uptake by a wild type rice (cv Oochikara) and a mutant (GR1) defective in Si uptake. The uptake experiment was conducted in a nutrient solution containing 0.75 mM Si in the presence or absence of inhibitors (1mM for DNP and 10 mM for NaCN) for 6 h. For low temperature treatment the plants were exposed to 4°C.

phenotype between the wild type (WT) and GR1. The short-term and relatively long-term uptake experiments showed that Si uptake by the roots in GR1 was significantly lower than that in WT at either a low or high Si concentration (Figure 6.12). However, there was no difference in the uptake of other nutrients such as P and K (Table 6.13). When the external solution contained 0.15mM Si, the Si concentration in the xylem sap of WT was 33-times as high as that in the external solution, whereas that of GR1 was only 3-times as high. The uptake of Si by WT was inhibited by metabolic inhibitors such as NaCN, 2,4-dinitrophenol and low temperatures, while the Si uptake by GR1 was not inhibited (Figure 6.13). These results suggest that the active transport system for Si uptake is defective in GR1. In the population of F_2 between GR1 and WT, the ratio of the plants with a high ability to take up Si to those with a low ability was 3:1, suggesting that the low ability to take up Si in GR1 is controlled by a recessive gene. This mutant is expected to be a powerful tool for isolation and identification of the Si transporter in rice roots.

6.6. SIMILAR MODE OF UPTAKE FOR SILICON AND GERMANIUM

Germanium (Ge) is a cognate element of Si and has chemical properties similar to those of Si. The response of plants to Ge was compared with that to Si in a series of studies (Takahashi et al., 1976a, b, c).

6.6.1. Effect of Ge on the growth

Several plants having different capacity for Si uptake were cultured in a nutrient solution containing 47 ppm silicic acid and 1-10 ppm germanic acid. After two weeks, their growth was compared (Takahashi et al., 1976a). Although the growth of all plant species was inhibited by exposure to Ge, the degree of inhibition varied widely with the plant species, and was in the order of rice>maize・cucumber・kidney bean>tomato>morning glory・ bindweed. Plant growth was inhibited by Ge most severely in rice and necrosis spots were observed in the rice leaf blades (Figure 6.14, Ma, 2002a). The resistance to Ge is negatively correlated with the capacity to take up Ge (namely Ge concentration in the top) and negatively correlated with the capacity to take up Si. The uptake of Ge was suppressed by the presence of Si and the Ge-induced inhibition of growth was alleviated in plants with a high capacity to take up Si.

Similar results were also obtained in an experiment with soil culture (Takahashi et al., 1976b). Four species (rice, maize, kidney bean, and oat) were grown in a soil amended with $CaCO_3$ or $CaSiO_3$ in the presence of Ge (GeO_2 powder, 50 and 100 ppm) or absence of Ge. The pH (H_2O) of the soil was increased from 4.8 to 5.6 by application of $CaCO_3$ and $CaSiO_3$. The growth of all species was inhibited by the addition of Ge, but the extent of inhibition

Table 6.14
Alleviative effect of Si on Ge toxicity in different plants

Species*	GeO_2 (ppm) added	Top dry weight (g/pot)		Top Si content (%)		Top Ge content (ppm)	
		$CaCO_3$	$CaSiO_3$	$CaCO_3$	$CaSiO_3$	$CaCO_3$	$CaSiO_3$
Rice	0	20.61	21.86	0.71	2.06	0	0
(water logged)	50	0.27	3.36	2.09	4.41	11750	3500
	100	0.09	1.48	7.20	5.63	14982	7170
Rice	0	18.99	20.58	0.48	1.40	0	0
(upland)	50	0.52	3.69	2.39	3.57	2870	1580
	100	0.11	1.61	4.79	5.42	11390	5540
Maize	0	25.53	27.53	0.37	0.61	0	0
	50	12.20	18.05	0.35	0.69	860	480
	100	4.32	5.97	0.70	1.18	3060	2470
Kidney bean	0	19.80	20.60	0.38	0.51	0	0
	50	8.43	12.63	0.40	0.90	760	430
	100	3.59	5.46	0.40	0.70	1430	700
Oat	0	17.36	17.59	0.24	0.75	0	0
	50	12.17	13.24	0.42	0.96	840	730
	100	5.31	7.95	0.51	0.96	2030	1040

* Culture period was 6 weeks for rice, 4 weeks for maize and kidney bean, and 5 weeks for oat.

Figure 6.14. Brown spots on the rice leaf blades cause by Ge. A, normal leaf; B, Ge-treated leaf

varied greatly with the plant species, which was in the order of rice (water logged)>rice (upland)>maize, kidney bean>oat (Table 6.14). This inhibition degree is consistent with the Ge or Si uptake capacity. The Ge-induced inhibition was alleviated by addition of Si in all species, but the alleviative effect was more obvious in the species with a high Si uptake capcity. These results suggest that although the physiological effects of Ge are different from those of Si, the uptake of Si by the roots is similar to that of Ge.

A

B

C

1978-5-1
播種後1ヶ月
GeO2添加後12日
+GeO2
−GeO2

D

1978-5-1
播種後1ヶ月
GeO2添加後12日
+GeO2
−GeO2

E

Figure 6.15. Growth of Italian ryegrass (A), rice (B), maize (C), red clover (D), and alfalfa (E) affected by Ge. Seedlings of 2-week old grown on vermiculite were treated with 10 ppm Ge for 12 days.

Figure 6.15 shows the growth of Italian ryegrass, rice, maize, red clover, and alfalfa grown on the vermiculite with or without 10 ppm Ge for 12 days (Takahashi et al., 1978). The growth of Italian ryegrass, rice, and maize was significantly inhibited by Ge, while that of red clover and alfalfa was hardly inhibited. These results indicate that Ge can be used to measure Si uptake ability.

Table 6.15
Ge uptake by intact plant and excised shoot

Species	Amount of ^{68}Ge in the shoot (^{68}Ge cpm/mg fresh wt.)		
	Intact plant (A)	Excised shoot (B)	A/B
Rice	2330.0	90.0	26.0
Kidney bean	29.2	23.0	1.3
Bindweed	6.6	34.4	0.2
Morning-glory	4.4	22.4	0.2

6.6.2. Similarity in uptake between Si and Ge

The mechanism of Ge uptake was further examined using a radioisotope of Ge (^{68}Ge) with a half-life of 282 days (Takahashi et al., 1976). The role of roots in Ge uptake was investigated in several plant species including rice, kidney bean, bindweed, and morning glory, which have different Si uptake capacity. When the roots of rice were cut off, the uptake by the excised shoots was only 1/26 of that by the intact plants (Table 6.15). However, the uptake of Ge by kidney bean was hardly changed by cutting off the roots and the uptake by bindweed and morning glory was increased 5 times by cutting off the roots.

The concentration of Ge in the xylem sap was seven times higher than that in the external solution at 4 hour after the application of Ge and increased to 70-fold at the 32nd hour (Table 6.16). The uptake of Ge by rice was strongly inhibited by metabolic inhibitors such as DNP (dinitrophenol), 2,4-D, and less inhibited by NaCN (Figure 6.16). The inhibitory effect of Na-malonate on the Ge uptake was not observed in rice. By contrast, the presence of metabolic inhibitors increased the uptake of Ge at the initial period of uptake morning glory (Figure 6.16). The effect of transpiration on the Ge uptake was not observed during a short time in rice, but the radio-autograph showed that the distribution of Ge in the shoot was controlled by the transpiration. The Ge-induced necrosis spots were located at the site of ^{68}Ge accumulation on the radio autograph, suggesting that the necrosis spots are caused by Ge accumulation.

The Ge-induced inhibition of the growth was alleviated by the presence of Si in rice. The uptake of ^{68}Ge by rice decreased with increasing Si concentration in the solution (Table 6.17). The content of Ge in the roots was hardly affected by the presence of Si, while that in the shoot was significantly decreased. This result suggests that the alleviative effect of Si on Ge-induced inhibition of the growth is attributed to decreased uptake, translocation and accumulation of Ge to the top by Si.

Table 6.16
Ge concentration in the bleeding sap of rice

Time after application of Ge (h)	Ge concentration (ppm) in medium (A)	Ge concentration (ppm) in bleeding sap (B)	B/A
0	5.0	–	–
4	4.4	33	7
24	2.3	106	47
32	1.9	136	71

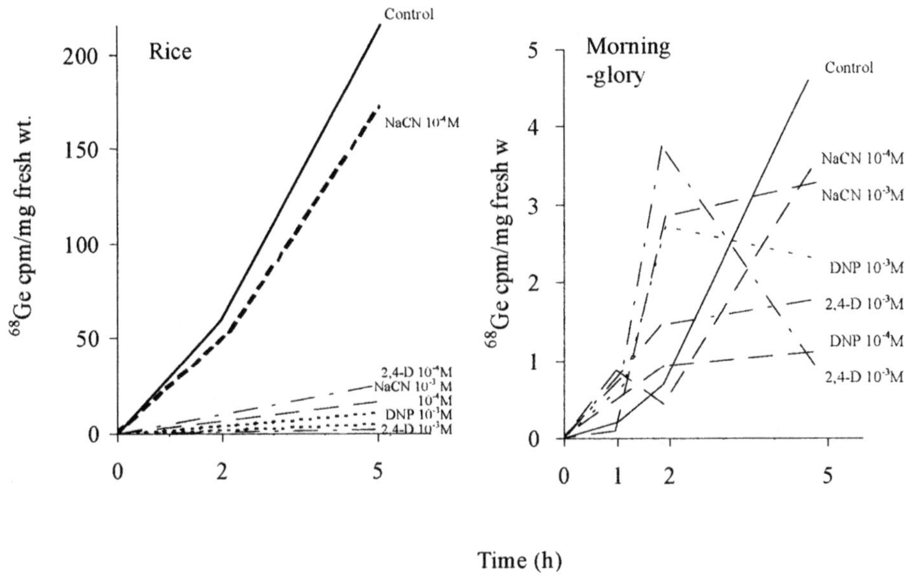

Figure 6.16. Effects of metabolic inhibitors on ^{68}Ge uptake by rice and morning-glory.

The radioisotope is useful in the studies on the behavior of an element in plants. However, the radioisotope of Si has a half-life of only 2.62 h, which is too short to be used. The results of the studies mentioned above indicate that Ge is similar to Si in terms of uptake. Thus, the radioisotope of Ge could be useful for the studies on Si uptake mechanisms.

Table 6.17
Effect of Si on Ge uptake in rice*

^{68}Ge concentration	Si concentration (ppm) in the solution				
(cpm/mg dry wt.)	0	5	25	50	100
Top	160	147	78	57	34
Root	12	12	11	11	12
Whole plant	124	115	63	47	26
Top/root ratio	13.3	11.8	7.0	5.2	1.3

*Ge was applied at 5 ppm to the solution

6.7. CHEMICAL FORM AND ACCUMULATION PROCESS OF SILICON IN RICE

The high Si content in the rice shoot suggests that Si after uptake is immediately translocated to the shoot. The chemical forms and distribution of Si in rice tissues were intensively investigated by Yoshida et al. (1959a, 1962d). Based on the reactivity of Si with molybdate, they fractionated Si in the rice plant into low molecular weight Si, colloidal Si and Si gel. Low molecular weight Si is a form which reacts with ammonium molybdate and mainly consists of orthosilicic acid. Colloidal Si is a form of Si sol which dissolves or dispers in water, but does not react with ammonium molybdate. Silica gel is a form not soluble in water. The results indicate that the low molecular weight Si accounted for less than 10% of total Si, colloidal Si was less than low molecular weight Si, and most Si was present in the form of Si gel. Furthermore, they found that Si in rice is mainly present as an amorphous form and that the possibility of Si in the form of organic Si compound is small based

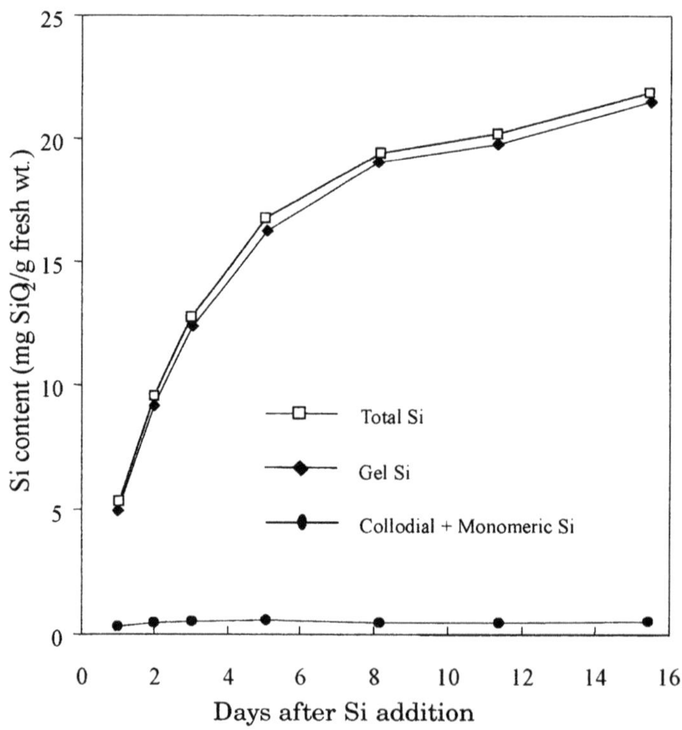

Figure 6.17. Different forms of Si in the shoots of rice cultured in a solution containing 100 ppm silicic acid.

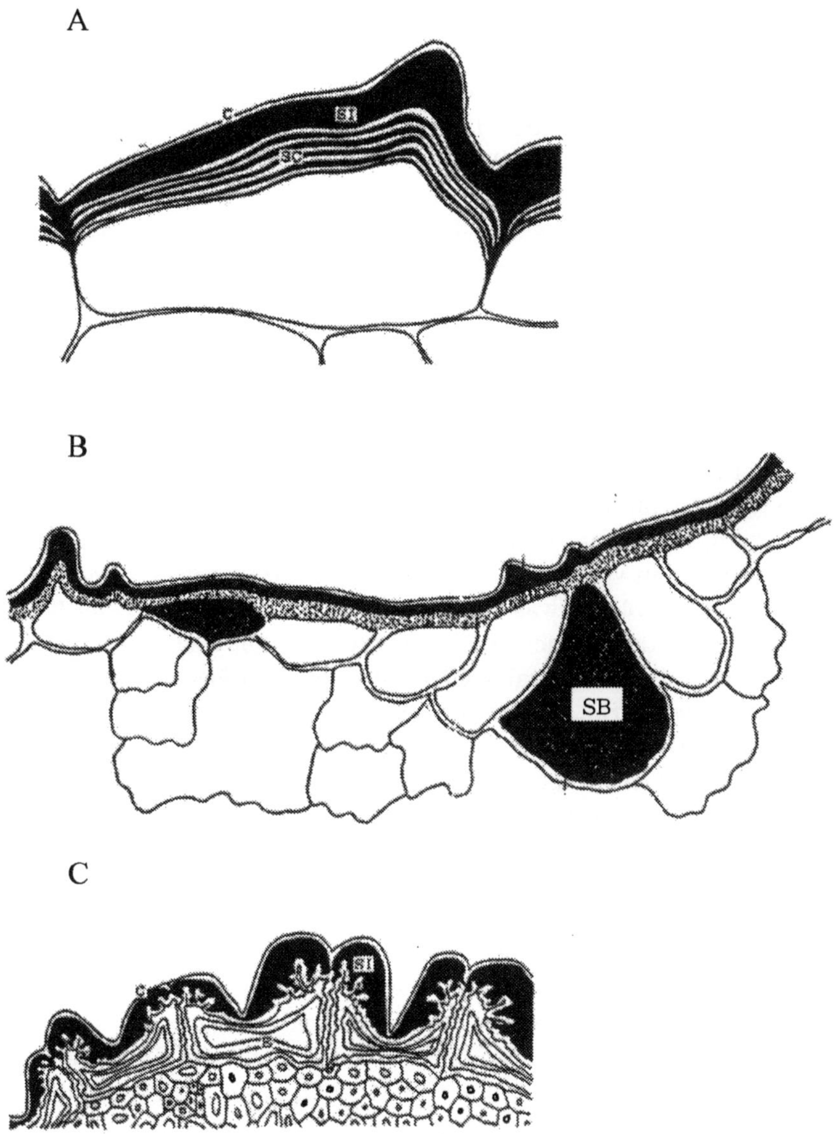

Figure 6.18. Diagram of Si deposition in different tissues of rice. A, cuticle-silica double layer in the leaf blade; B, silicified cells; C, cuticle-silica double layer in hull of rice grain. Blackened area represents location of Si deposition. C, cuticle; SI, silica layer; E, epidermis; SC, silica cellulose membrane; SB, silica body.

on infrared absorption spectra. From the dissolution test in hot water, Si in the rice plants was found to exist in a form similar to silica gel.

When cultured in a nutrient solution containing 100 ppm silicic acid, the total Si content in the rice shoots increased with the duration of Si supply, from 2.5% SiO_2 (dry weight basis) on the first day after silicic acid supply to 14% SiO_2 on the 15th day (Ma, 1990, Figure 6.17). At any time, the content of Si gel always accounted for more than 90% of total Si and the content of colloidal Si and monomeric Si was kept between 300 and 500 ppm SiO_2. Silicic acid polymerizes at a concentration exceeding 2 mM at 25°C. These results indicate that after uptake, silicic acid is rapidly translocated to the top along with the transpiration steam, and with loss of water due to transpiration, silicic acid is gradually concentrated and polymerized to colloidal, and finally to gel form (silica).

To observe the localization of Si in rice tissues, Yoshida (1965) developed an HF etching method by embedding the tissues in a meta acryl resin and then incubating with HF. Most Si was localized in the epidermis of leaf blades and hulls of rice grain, where the transpiration rate was high. In the roots, Si was distributed uniformly in various tissues. Electron microscopy showed that the cuticle layer of rice leaf blades was very thin (0.1 μm). A thick Si layer (2.5 μm) was observed beneath the cuticle and several layers of cell walls were filled with Si, and the whole cells of several epidermal cells (bulliform cells, long short cells) were filled with Si. Since the Si layers beneath the cuticle are characteristic of Si-accumulating plants, these layers are referred to as cuticle-silica double layers (Figure 6.18). These layers play an important role in controlling transpiration and resistance to diseases and pests as discussed below.

The change of Si forms with growth progression and the silicification process of epidermal cells were also investigated by Ma (1990). Two types of silicified cells which are termed as silica cells and silica bodies or silica bulliform cells, were observed in the leaf blades supplied with Si. Silica cells are located on vascular bundles, showing a dumbbell-shape, while silica bodies are in bulliform cells of rice leaves (Figure 6.19, Ma, 1990). In the shoot containing less than 5% SiO_2, silica bodies were hardly observed, though silica cells were observed. With increasing Si content, silica bodies increased. These results suggest that the silicification of epidermal cells proceeds from silica cells to silica bodies (Figure 6.19). A good correlation was observed between Si content of rice shoot and the number of Si bodies (Figure 6.20). Takeoka et al. (1983) developed a method for observation of silica bodies on the leaf blades using soft x-ray. Figure 6.21 shows silica bodies in the blade-leaf of rice with different Si content (Ma, 1990).

A

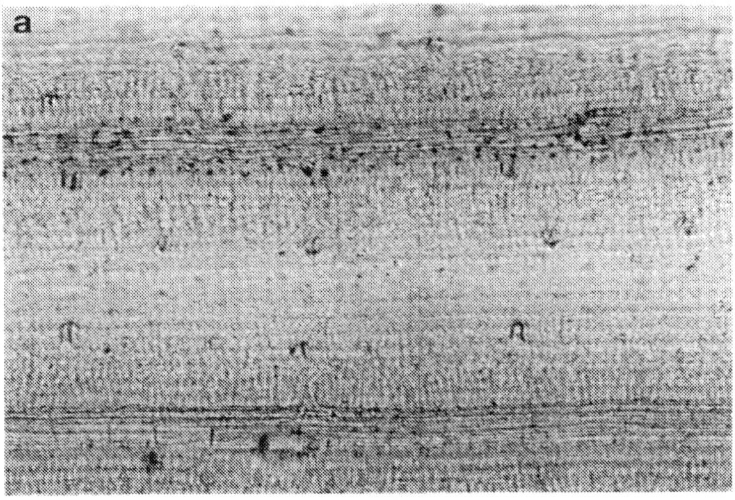

B

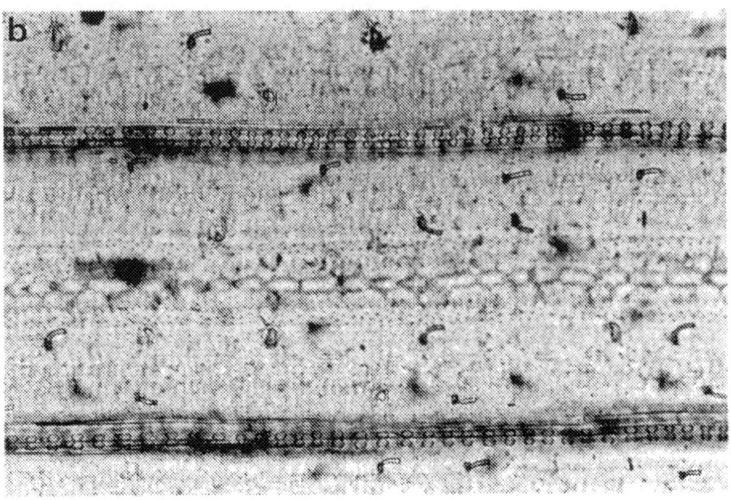

C

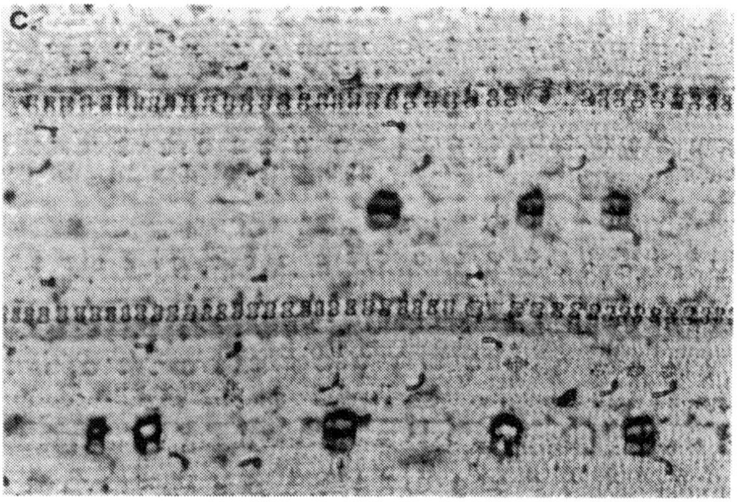

D

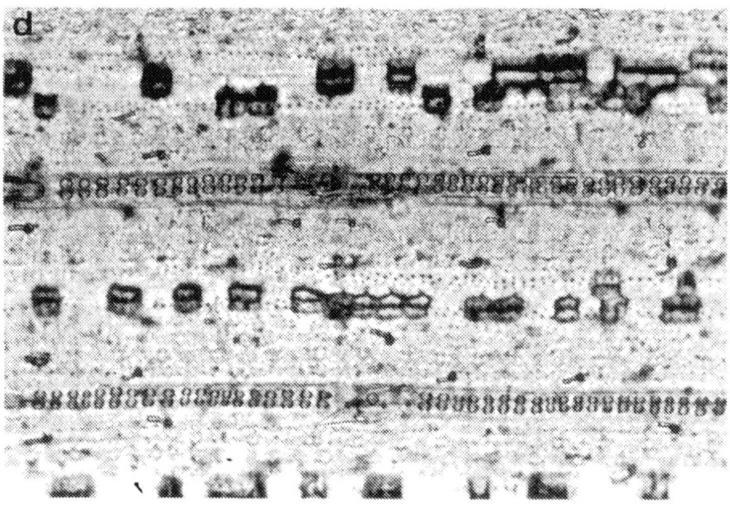

Figure 6.19. Formation process of silicified cells in rice leaf blades. The region between 8 and 10 cm from the tip was examined under a microscope. A, the Si content of the shoot is 0.11% of SiO_2, showing no silicified cells; B, the Si content is 4.07% SiO_2, showing silica cells on the vein; C, the content is 8.75% SiO_2, showing silica cells and few silica bodies; and D, the content is 10.5% SiO_2, showing many silica bodies.

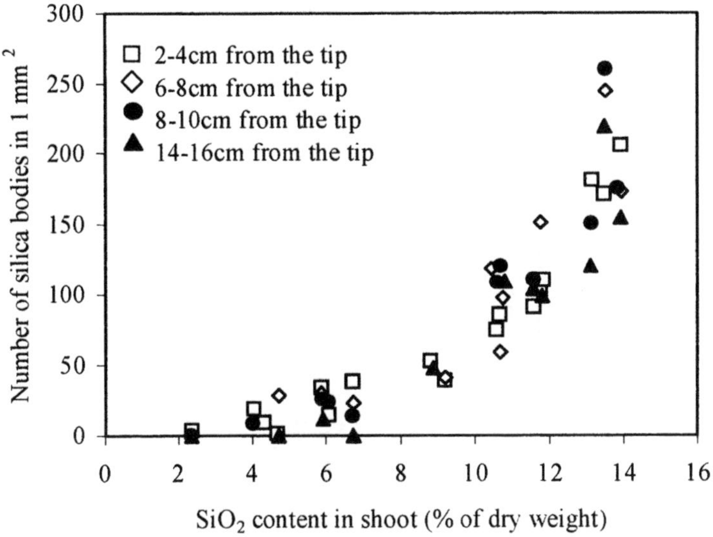

Figure 6.20. Relationship between Si content in the rice shoot and the number of silica bodies at various parts of the third leaf blades.

A

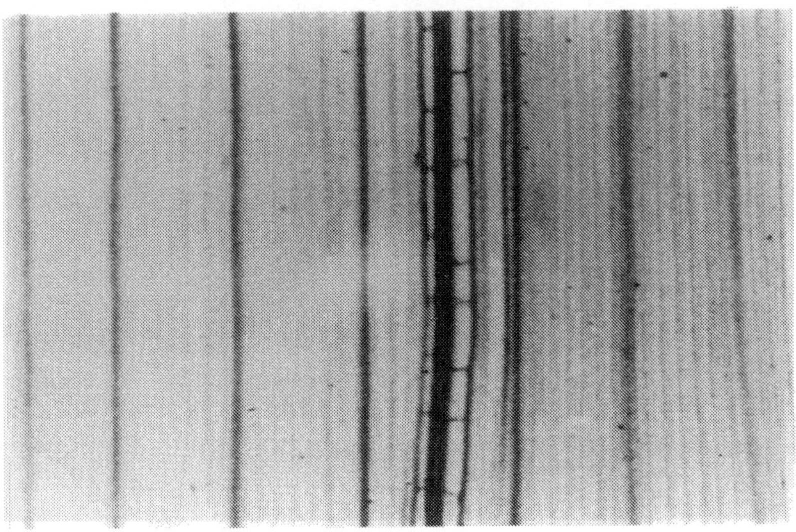

B

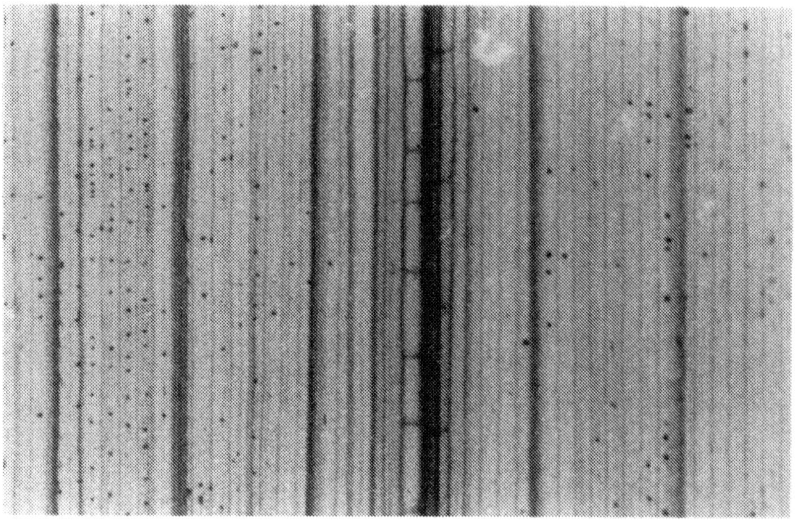

C

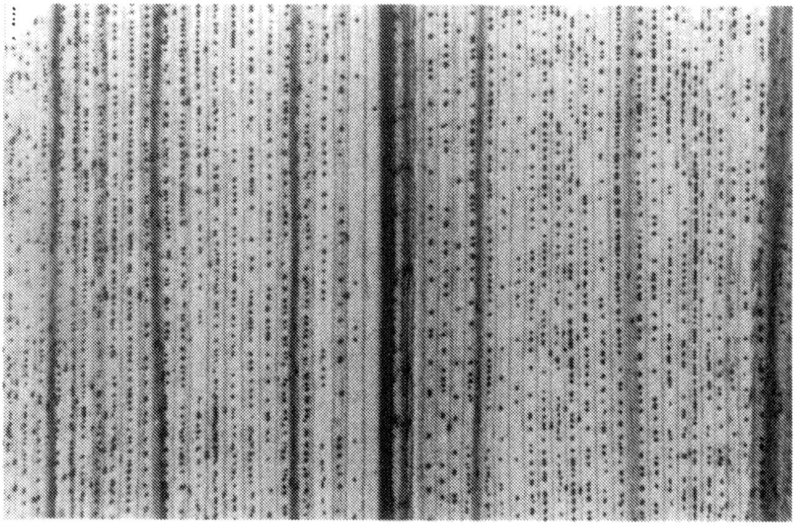

Figure 6.21. Silica bodies detected by soft x-ray. A, no silica body; B, few silica bodies, and C, many silica bodies.

In addition to leaf blades, silicified cells were also observed in the epidermis and vascular tissues of stem, leaf sheath and hull. This silicification enhances the strength and rigidity of cell wall, thus increasing the resistance of rice plants to diseases and lodging, improving light-receiving plant form in a community, and decreasing transpiration.

Chapter 7

Functions of silicon in plant growth

Although all plants rooting in soil contain Si in their tissues, earlier researchers have scarcely studied the how Si affects plant growth. This is mainly because Si is so abundant in soil that Si deficiency hardly occurs in most plants. Furthermore, the symptoms of Si deficiency are not as visible as those of the deficiency of other essential elements. However, at the beginning of the twentieth century, Japanese scientists realized that Si is important for the production of rice and since then a number of studies have been carried out on the effects of Si on plant growth, especially on rice, which is the most important crop in Japan. In this chapter, results on Si functions in plant growth are summarized.

7.1. BENEFICIAL EFFECTS OF SILICON ON PLANT GROWTH

Water culture is a valuable technique for the studies on the effect of Si on plant growth because Si is easily removed from the growth medium. The beneficial effects of Si on plant growth have been observed in a wide variety of plant species. The effect of Si varies with the plant species; and usually is more prominent in the plants accumulating a large amount of Si. In this section, the beneficial effects of Si in various plant species are described.

7.1.1. Rice

7.1.1.1. Deficiency symptoms

Rice is a typical Si-accumulator and the Si content in the shoot reaches up to 10%, which is several-fold higher than the contents of macronutrients such as N, P, and K. The effect of Si deficiency on rice growth has been investigated by many researchers. Ohkawa (1936) supplied colloidal Si which was prepared by neutralizing water glass and then dialyzing, to rice in a nutrient solution and observed that rice supplied with Si had erect leaves and stems. The dry weight of the shoot and grain yield was increased by Si supply. Si deficiency increased the number of empty grains, resulting in white and soft heads. He wrote that such white heads resembled the head of Japanese pampas-grass (*Miscanthus sinensis*).

Okamoto (1956, 1957a,b) reported that Si deficiency inhibited rice growth and promoted dying-off of leaf blades after heading, leading to the difference in the dry weight between the rice with and without Si supply. He also observed dark brown spots on the stem and head of rice without Si supply and further isolated *Piricularia oryzae* from the spots. He concluded that Si may not be an essential element for plant growth because rice without Si can ripen, but Si is an agronomically essential element for rice because Si deficiency causes significant reduction of grain yield.

Yoshida et al. (1959b) found that the growth of rice was only slightly affected by the absence of Si from the culture solution during the vegetative growth stage. However, after around heading, Si deficiency increased the dying-off rapidly and growth was suddenly reduced the growth weight resulting in an unexpectedly low grain yield.

They also found that at a younger stage, the leaf-blade of the Si-deficient shoot was bent downwards just like a weeping willow and that in such plants, the transpiration rate at the time of the initiation of panicle primordia was 33% higher than that of rice with Si. From these phenomena they postulated that Si contributed to better growth of the rice through depression of excessive transpiration, because excessive transpiration causes physiological shortage of water around heading, a time at which the rice plant requires vigorous transpiration while the functions of rice roots begin to decline.

Okuda and Takahashi (1961a, b) cultured rice by Si-free culture. Water used for preparation of nutrient solution was produced by distillation with a tin-lined copper still followed by passing through the column of cation-exchange resin to remove trace amount of copper dissolved from the still. Refined chemicals were used, resulting in Si concentration of less than 0.002 ppm in the nutrient solution for rice culture.

Silicon was given in the form of silicic acid, which was prepared by passing 500 to 1000 ppm Si of sodium silicate solution through a H-type cation exchange-resin column. Instead of glass wares, plastic wares and polyethylene containers were used. Polystyrol pellet (as synthetic sand for nursery), polyester sponge mat for fixing the plants were also used. Cultivation was done in the open air attending to keep free of disease and fine dust contamination.

Under such conditions described above, Okuda and Takahashi (1961a, b) found that Si greatly influenced the growth of rice. At the early tillering stage, the leaf blades of –Si plants were noticeably drooped for some weeks and then gradually recovered (Figure 7.1). Silicon supply, especially at the stage of reproductive growth, increased dry weights of leaves and stems measurably. After heading, the dry weight of years increased rapidly as the ripening progressed in the plants with Si supplied (+ Si plants), but not in the –Si plants.

At harvest time, the dry weight of ears of –Si plants (containing 0.06% Si in the leave) was about half of that of +Si plants (containing 8.39% Si).

A

B

C

D

Figure 7.1. Symptoms of Si deficiency in the rice at the tillering stage. The plants grown with or without Si for 23 (A), 39 (B), 43 (C), and 62 (D) days are shown.

Although the growth of rice was greatly influenced by Si, +Si plants and –Si plants did not suffer from diseases and the characteristic symptoms of Si deficiency described by Wagner (1940) such as reddish brown spots on leaves

A

B

Figure 7.2. Symptoms of Si deficiency in rice at a young stage (A) and ripening stage (B). The plant with Si was supplied with 100 ppm SiO_2 as silicic acid.

were not observed, but the lower leaves at the ripening stage withered earlier in the –Si plants than in the +Si plants.

A typical symptom of Si-deficiency in rice at the young stage and ripening stage is shown in Figure 7.2 (Ma, 1998).

7.1.1.2. Effect of time of Si supply on the growth and grain yield

Okuda and Takahashi (1961b) investigated the effect of Si on the growth and yield of rice at various growth stages. Four treatments were designed; Si was not supplied (-Si-Si) or supplied (+Si+Si) during whole growth period, or Si was supplied only before (+Si-Si) or after (-Si+Si) panicle formation stage.

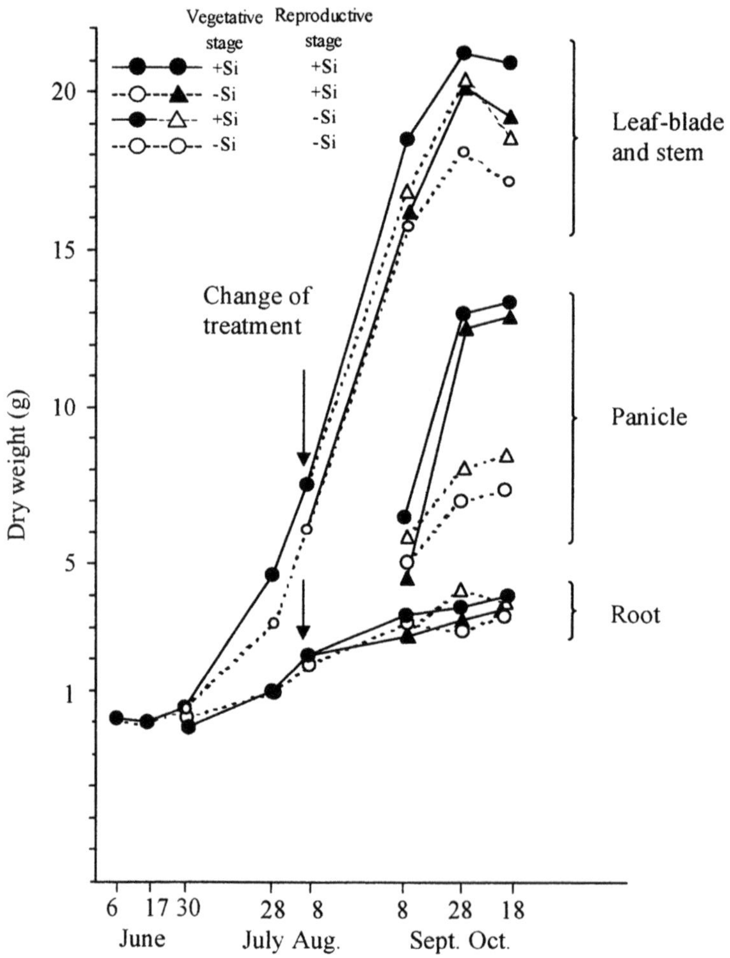

Figure 7.3. Effect of Si supply at various growth stages on the growth of rice.

Table 7.1
Effect of Si supply at vegetative and/or reproductive growth stage on the rice grain yield and its components

	Treatment			
	-Si-Si	+Si-Si	-Si+Si	+Si+Si
SiO_2 content (%)				
Top	0.05	2.16	6.88	10.41
Root	0.02	0.45	3.38	3.42
Total dry wt. (g/pot)	27.5	30.8	35.2	38.3
Yield components				
Number of panicles	9.5	10.3	10.0	11.0
Number of spikelets per panicle	49.3	47.1	65.4	63.2
Percentage of ripening	55	67	78	76
Weight of 1000 mature kernels (g)	20.4	20.4	20.2	20.5
Weight of winnowed paddy (g/pot)	5.25	6.64	10.30	10.83

Silicon was supplied as silicic acid at 100 ppm SiO_2 and all plants were not attacked by diseases and pests. Compared with the Si supply during vegetative growth stage, the Si supply during the reproductive growth stage was more effective in increasing plant growth and grain yield (Figure 7.3, Table 7.1). Silicon supply only during reproductive growth stage produced grain yield comparable to that during whole growth period.

Among the four yield components, the number of panicles was determined mainly during the vegetative growth stage, while the number of spikelets per panicle, the percentage of ripened grains, and the weight of 1000 kernels were determined mainly during the reproductive growth stage. Silicon deficiency during only the vegetative growth stage decreased the number of panicles and that during the reproductive growth stage decreased the the number of spikelets per panicle and the percentage of ripened grains. The length of panicles was shortened by Si deficiency during the reproductive growth stage. Thus, the grain yield of –Si-Si plants and +Si-Si plants were nearly half of those of +Si+Si plants and –Si+Si plants.

Ma et al. (1989) further subdivided the growth period of rice into vegetative (from transplanting to panicle initiation), reproductive (from panicle initiation to heading) and ripening (from heading to maturity) stages and investigated the effect of Si addition or removal during each stage on the plant growth, Si uptake and distribution of Si (Figure 7.4). When Si was removed during the reproductive stage, the dry weights of straw and grain decreased by 20 and 50%

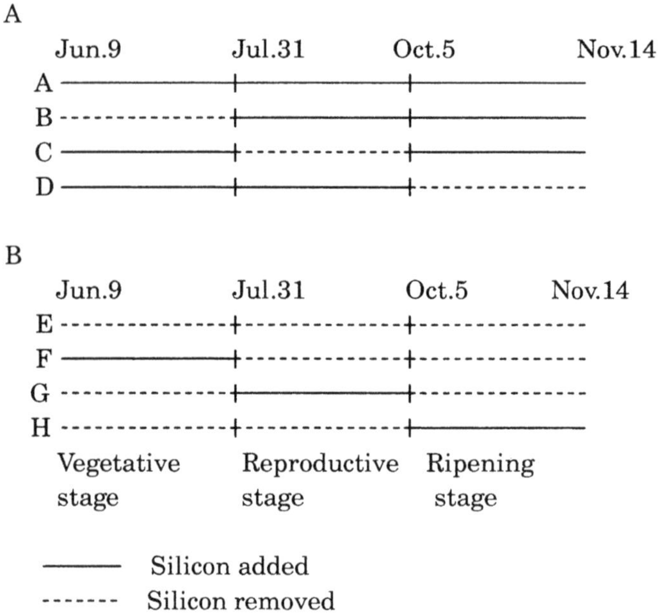

Figure 7.4. Design of experiment for Si supply at various growth stages. A, Si removal experiment; B, Si addition experiment.

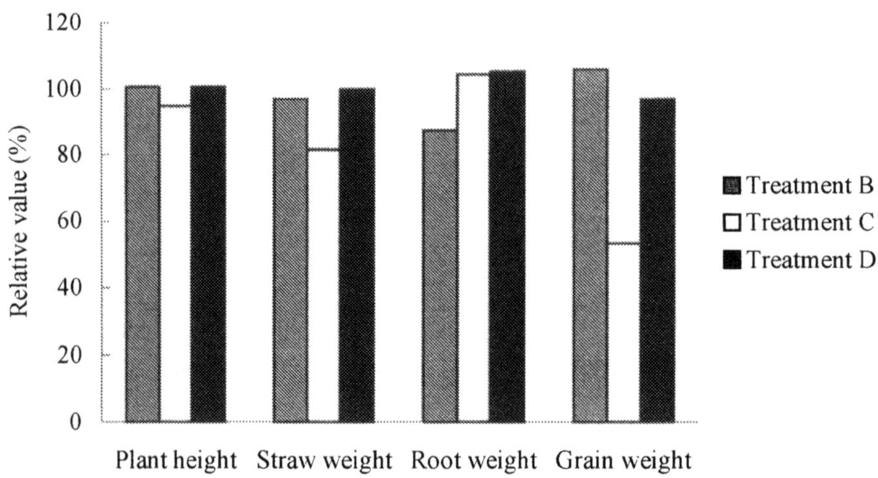

Figure 7.5. Effect of the removal of silicon at various growth stages on the growth of the rice plant. Values in treatment A as 100, plant height, dry weights of straw, root, and grain in Treatment A were 105cm, 86.3, 10.1, and 20.1g/pot, respectively.

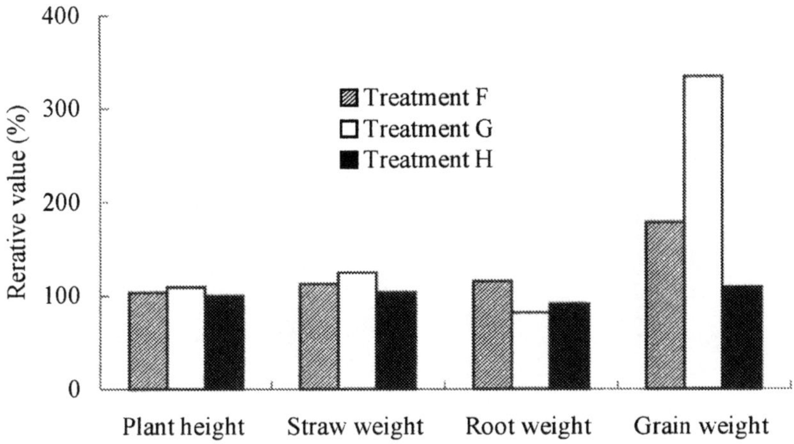

Figure 7.6. Effect of the addition of silicon at various growth stages on the growth of the rice plant. Values in treatment E as 100, plant height, dry weights of straw, root, and grain of Treatment E were 91.3cm, 61.5, 9.3, and 4.2g/pot, respectively.

respectively, compared with those of plants cultured in the solution with Si throughout the growth period (Figure 7.5). Removal of Si during vegetative or ripening stage hardly affected the growth. In agreement with this result, when Si was added during the reproductive stage, the dry weights of straw and grain increased by 30 and 243%, respectively, over those of the plants cultured in a Si-free solution throughout the growth period (Figure 7.6). Addition of Si during vegetative stage also increased grain weight, but the addition of Si during the ripening stage had little effect.

The percentage of filled spikelets was markedly increased by the addition of Si during the reproductive stage, and was markedly decreased by removal of Si during that stage (Figure 7.7A, B). The 1,000-grain weight was hardly influenced by the addition or removal of Si regardless of the growth stage.

The uptake percentage of Si during the vegetative, reproductive and ripening stages was about 10, 65, and 25%, respectively, in both removal and addition experiments (Tables 7.2, 7.3). About 70 to 75% of Si in the leaf blades were absorbed during the reproductive stage, while about 75% of Si in the panicle was absorbed during the ripening stage. Forty to 50% of the Si absorbed during the vegetative and reproductive stages was present in the leaf blades (Tables 7.4), whereas only 20 to 30% of Si absorbed during the ripening stage was present in the leaf blades. More than 99% of Si absorbed was distributed in the shoot.

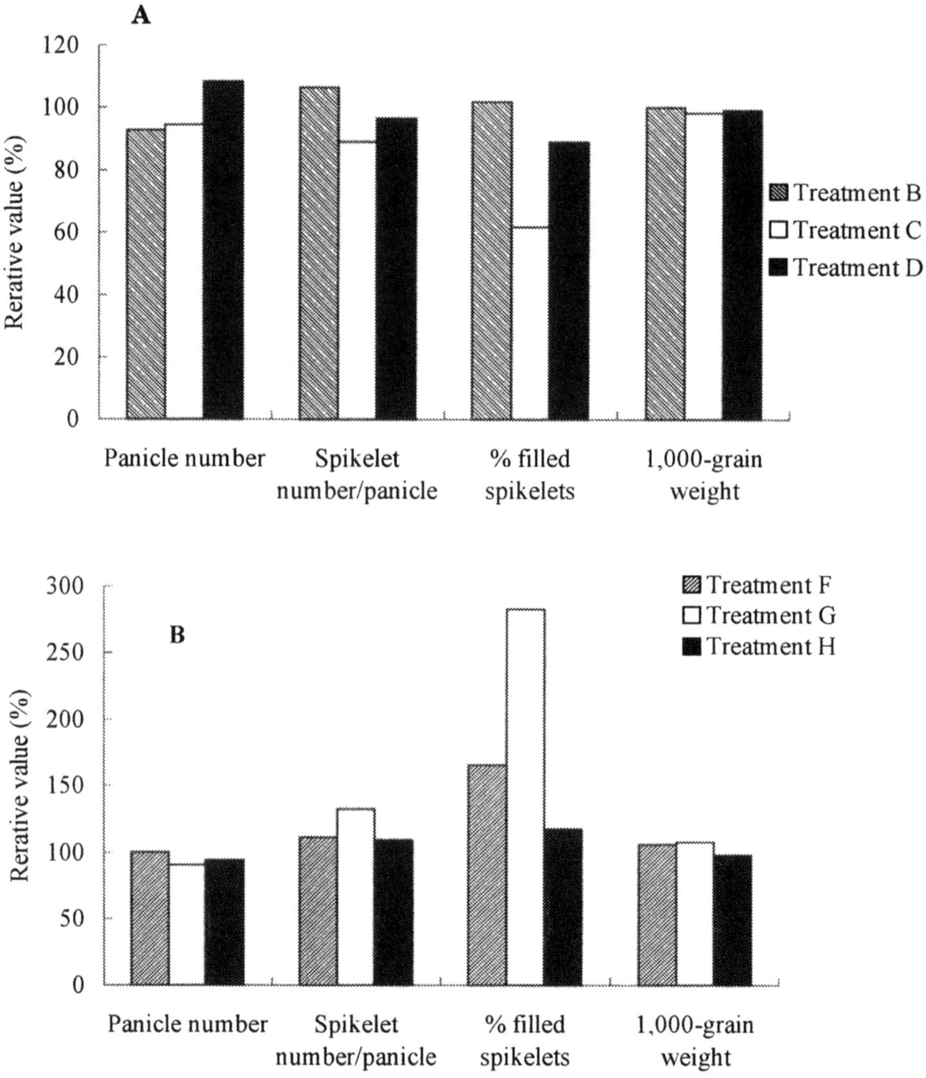

Figure 7.7. Effect of Si removal (A) or addition (B) at various growth stages on the yield components of rice plant. In A, values in treatment A as 100, the panicle number per pot, spikelet number per panicle, % filled spikelets, and 1,000-grain weight in treatment A were 23, 75, 54 and 21.9g, respectively. In B, values in treatment E as 100, the panicle number per pot, spikelet number per panicle, % filled spikelets, and 1,000-grain weight in treatment E were 22, 57, 16, and 21g, respectively.

Table 7.2
Uptake of silicon by the rice plant during various growth stages (Si removal experiment).

Plant part	Total Si uptake ($mg SiO_2$/pot)	Uptake percentage during		
		Vegetative stage	Reproductive stage	Ripening stage
Panicle	920	0.0	24.0	76.0
Leaf blade	4820	10.4	75.4	14.3
Stem	7590	11.5	65.2	23.3
Shoot	13330	9.6	66.5	23.8
Root	84	0.0	69.7	30.3
Total	13414	9.6	66.5	23.8

Table 7.5 shows the Si content of each organ of rice with various treatments. Removal or addition of Si during the reproductive stage especially affected the Si content of the flag leaf, whose photosynthesis contributes significantly to the yield.

These results clearly indicate that the supply of Si during the reproductive stage is most important for plant growth.

Table 7.3
Uptake of silicon by the rice plant during various growth stages (Si addition experiment).

Plant part	Total Si uptake ($mg SiO_2$/pot)	Uptake percentage during		
		Vegetative stage	Reproductive stage	Ripening stage
Panicle	770	1.7	23.5	74.8
Leaf blade	5930	12.3	70.2	17.6
Stem	6840	7.3	65.9	26.8
Shoot	13540	9.2	65.4	25.5
Root	76	6.5	47.4	46.1
Total	13616	9.1	65.3	25.6

Table 7.4
Distribution of Si in various plant parts during growth stage

Stage	Distribution percentage in				
	Panicle	Leaf blade	Stem	Shoot	Root
Removal experiment					
Vegetative	0.0	36.4	63.6	100	0.0
Reproductive	2.5	41.0	55.9	99.4	0.7
Ripening	22.0	21.6	55.6	99.2	0.8
Addition experiment					
Vegetative	1.1	58.4	40.1	99.6	0.4
Reproductive	2.0	46.8	50.8	99.6	0.4
Ripening	16.6	29.9	52.5	99.0	1.0

7.1.1.3. Effect of Si supply levels on the growth and grain yield

Okuda and Takahashi (1961c) examined the effect of Si supplied at various concentrations (0, 5, 20, 60, and 100 ppm SiO_2) on the growth and grain yield of the rice. At the early tillering stage, the leaf-blades of rice supplied with 0 and 5 ppm SiO_2 were visibly droopy for some weeks. The top length, number of stems, dry weight and grain yield increased with increasing Si levels (Table 7.6). Among yield components, the percentage of ripening was most affected by Si

Table 7.5
Silicon content in plant parts of rice with various treatments

Treatment	Stage at which Si was removed or Added	SiO_2 content (%)				
		Panicle	Flag leaf	Leaf blade	Stem	Root
A	No removal	3.24	10.9	20.4	11.8	0.93
B	Vegetative	3.43	12.0	18.1	11.1	1.33
C	Reproductive	3.53	5.19	6.73	4.59	0.34
D	Ripening	0.77	9.42	17.2	9.30	0.66
E	No addition	0.02	0.05	0.20	0.11	0.04
F	Vegetative	0.07	0.20	3.43	1.00	0.04
G	Reproductive	0.77	9.99	17.6	8.20	0.45
H	Ripening	3.11	5.20	4.86	4.06	0.39

Table 7.6
Effect of various Si levels on the growth and yield of rice

	Si concentration supplied (SiO_2 ppm)				
	0	5	20	60	100
Top Si content (SiO_2 %)	0.07	0.62	2.00	5.19	8.01
Dry weight at harvest (g/pot)					
Panicle	4.3	5.1	6.0	7.6	9.8
Leaf and stem	13.1	14.5	15.1	15.0	14.8
Root	5.0	5.2	5.0	5.0	5.1
Date of heading	Aug.25	Aug.22	Aug.16	Aug.14	Aug.14
Yield components					
Number of panicles	11	12	13	12	13
Number of spikelets per panicle	50.7	48.5	53.5	52.3	51.4
Percentage of ripening	26.9	31.1	31.9	45.6	61.7
Weight of 1000 mature kernels (g)	19.3	20.5	20.3	21.0	20.9
Weight of winnowed paddy (g/pot)	2.9	3.7	4.5	6.3	8.6

supply. Silicon supply also hastened the time of heading. These effects of Si supply were noticeable at the concentration above 60 ppm SiO_2 (Table 7.6). These results suggest that the supply of a large amount of Si is necessary for healthy growth and high grain yield.

7.1.1.4. Effect of Si on the growth of various rice cultivars

Cultivars of rice in Japan have been bred for tolerance to high nitrogen and for heavy manuring and intensive cultivation. Miyake and Takahashi (1984) compared the effect of Si on the growth between the Japanese cultivar (Nihonbare) and Chinese cultivar (Tiehu-ai-1) and Indian cultivar (RP-9-5). As shown in Table 7.7 and Figure 7.8, Si had a significant effect on the yield in all cultivars. This indicates that the role of Si in rice growth and yield is identical in all cultivation areas.

7.1.1.5. Effect of Si on nutrient uptake

Yoshida et al. (1956) noted that the concentrations of N, P, Ca and Mn in the top of rice decreased by Si supply (Table 7.8), and they suggested from fractionation of calcium in leaf blades that calcium might partly replace Si in Si-deficient plant.

Table 7.7
Effect of Si on the yield of various rice cultivars

Cultivar	Weight of one panicle (g)	Number of grains per panicle	Percentage of ripened grain	SiO_2 % in leaf+stem
Japanese cultivar (Nihonbare)				
+Si	2.3	92	97	11.10
-Si	0.5	45	47	0.11
Chinese cultivar (Tiehu-ai-1)				
+Si	2.7	105	92	14.70
-Si	1.9	77	77	0.18
Indian cultivar (RP-9-5)				
+Si	3.4	147	87	11.50
-Si	1.7	100	79	0.13

A

B

C

Figure 7.8. Growth of various cultivars of rice supplied with or without Si. A, Japanese cultivar (Nihonbare); B, Chinese cultivar (Tiehu-ai-1); C, Indian cultivar (RP-9-5).

Table 7.8
Mineral composition of rice grown in the absence and presence of Si at harvest time

	Mineral content (%)						
	SiO_2	N	P_2O_5	K_2O	MgO	CaO	Mn
-Si							
Ear	0.05	1.88	1.10	1.00	0.30	0.06	tr.
Leaf-blade	0.36	1.33	0.52	0.68	0.69	0.75	0.021
Leaf sheath+stem	0.05	0.72	0.58	0.70	0.31	0.16	0.014
Root	0.07	1.11	0.53	-	0.09	0.07	tr.
+Si							
Ear	2.48	1.61	0.82	1.34	0.27	0.05	tr.
		(86)	(75)	(134)	(90)	(83)	
Leaf-blade	17.64	0.94	0.34	0.82	0.78	0.49	0.016
		(71)	(65)	(121)	(113)	(65)	(76)
Leaf sheath+stem	10.63	0.60	0.31	0.53	0.28	0.16	0.012
		(83)	(53)	(76)	(90)	(100)	(86)
Root	0.87	1.16	0.44	-	0.07	0.10	tr.
		(105)	(83)		(78)	(123)	

*Figures in parentheses are indices of +Si to -Si.

Okuda and Takahashi (1961c) investigated the effect of various Si concentrations on the nutrient uptake in rice. As the Si concentration in the nutrient solution increased, the contents of P, Fe and Mn in the top decreased gradually and this change was striking above the 60 ppm SiO_2 level (Table 7.9).

Ma and Takahashi (1989a) also found that the contents of all elements examined were decreased by the addition of Si, but the effect on the uptake varied with the element (Table 7.10). The uptake of N and K was not decreased by Si supply, but that of P, Ca, Fe, and Mn was markedly decreased. The mechanisms responsible for the effect of Si on the nutrient uptake remain to be investigated in the future.

Table 7.9
Effect of Si concentration on the mineral composition of rice at harvest time

SiO_2 concentration	Shoot mineral content (%)				
(ppm) supplied	N	P_2O_5	K_2O	Fe_2O_3	Mn_3O_4
0	1.41	0.43	1.10	0.044	0.010
5	1.28	0.39	1.09	0.039	0.008
20	1.24	0.36	1.05	0.038	0.008
60	1.16	0.34	1.02	0.029	0.005
100	1.12	0.30	0.91	0.026	0.005

Table 7.10
Effect of Si as silicic acid on nutrient uptake[a]

Element	Shoot content		Uptake amount (mg/pot)		+Si/-Si
	+Si	-Si	+Si	-Si	
N (%)	0.91	1.25	1051.2	947.7	1.11
P (%)	0.27	0.43	445.0	521.0	0.85
K (%)	1.19	1.66	1373.1	1258.0	1.09
Ca (%)	0.10	0.26	119.4	197.1	0.61
Mg (%)	0.23	0.37	267.0	283.2	0.94
Fe (ppm)	106.7	246.0	12.3	18.7	0.66
Mn (ppm)	192.6	439.9	22.1	33.4	0.66

[a] Rice was grown in a nutrient solution with or without 1.67 mM silicic acid until maturity.

7.1.2. Barley

In a long-term field experiment conducted at Rothamsted Experimental Station, the application of sodium silicate was found to increase the yield of

Table 7.11
Growth response of barley to silicic acid

		+Si	-Si
Dry weight (g/pot)			
Panicle		6.6	4.6
Stems+leaves		17.7	15.4
Roots		2.4	2.4
Ripened grains		5.9	2.1
Mineral content (%)			
Si	Top	1.00	0.01
	Root	0.90	0.01
N	Top	1.21	1.40
	Root	1.96	1.84
P	Top	0.18	0.22
	Root	0.34	0.33
K	Top	1.73	1.82
	Root	1.06	1.16

barley, but for an unknown reason. Okuda and Takahashi (1961d) investigated the effect of Si on the growth and yield of barley using water culture. In barley, the Si content of the shoots was much lower than that of rice (1%, Table 7.11). Silicon had little effect on the dry matter production, and uptake of P, N and K. However, similar to rice, the percentage of ripening of panicles was remarkably increased with addition of silicic acid, resulting in a big difference in the grain yield between the plants supplied with and without Si. The transpiration was slightly decreased by the addition of Si. The beneficial effect on the increased percentage of ripening may be attributed to the Si accumulated on the hull.

7.1.3. Tomato

Wooley (1957) cultured tomato in the absence of Si but did not observe any abnormal symptoms in the plant. The growth was nearly the same in the plants either with or without Si supply. Okuda et al. (1961d) also did not find any beneficial effect of Si in tomato although they found that tomato takes up Si rejectively in contrast to rice. However, Miyake et al. (1976b, c, d, 1978) reported symptoms of Si deficiency in tomato when they grew tomato until the flowering stage in water culture with strict Si deficiency and keeping the plant under the optimal growth conditions including air temperature, light intensity, day length and aeration.

Figure 7.9. Symptoms of Si deficiency in tomato (6 weeks after seeding).

Figure 7.10. Symptoms of Si deficiency in tomato at the flowering stage.

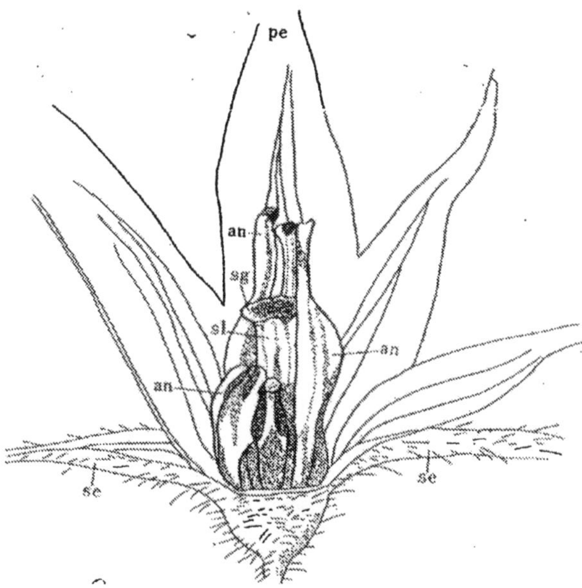

Figure 7.11. Abnormal flower with degenerated stamens (tomato plant cultured in Si-free solution). se, sepal; an, anther, sl, style; pe, petal; sg, stigma

Miyake et al (1976b) grew the plant very carefully in Shive & Robbins nutrient solution in the presence or absence of 100 ppm SiO_2 as silicic acid. The atmospheric temperature in the glasshouse was kept at 21-23°C in the day time and at about 18°C at night. The symptoms of Si deficiency appeared after the first bud flowered, and before that the growth was quite normal in the plant without Si supply. The symptoms of Si deficiency are characterized by that the growth of meristematic tissue at the top was markedly depressed, but the tissue was not blasted and that young leaves near the top were deformed, then became hardened and brittle (Figure 7.9). With the advance of deficiency, chlorosis developed on the upper leaves and necrosis spots appeared in the lower leaves spreading to the upper leaves. The plants without Si supply bloomed but failed to pollinated and bore malformed fruits or not fruits (Figure 7.10). The degeneration of stamens and abnormal shape of pollen-grain were observed (Figure 7.11). The pollen fertility was significantly decreased by Si deficiency (Table 7.12).

When Si was supplied to the plant already showing deficiency symptoms, new shoots developed about 3 weeks later showed a normal appearance (Figure 7.12), although the abnormal symptoms formerly formed remained unchanged. When the plants supplied with Si were subjected to the Si-free solution from the first bud flowering stage, the stem growth stopped and malformation of new leaves was observed after about 10 days (Figure 7.13). Since the plants had been supplied Si until the flowering stage, they bore some fruits, but brown spots were observed on the fruit skin.

The effect of Si supplied at various growth stages on the growth of tomato is shown in Table 7.13. Removal or addition of Si at the flowering stage has a significant effect on the growth and yield. These data are reflected in the growth pattern shown in Figure 7.14.

Table 7.12
Effect of Si deficiency on pollen fertility of tomato

SiO_2 ppm in solution	Time of pollen fertility tested	Fertility ratio (%)
0	just before bloom	82
0	in bloom	64
100	just before bloom	93
100	in bloom	91

Figure 7.12. Symptoms of Si deficiency in tomato plants that had not been supplied Si before flowering, but supplied Si thereafter (-Si+Si treatment).

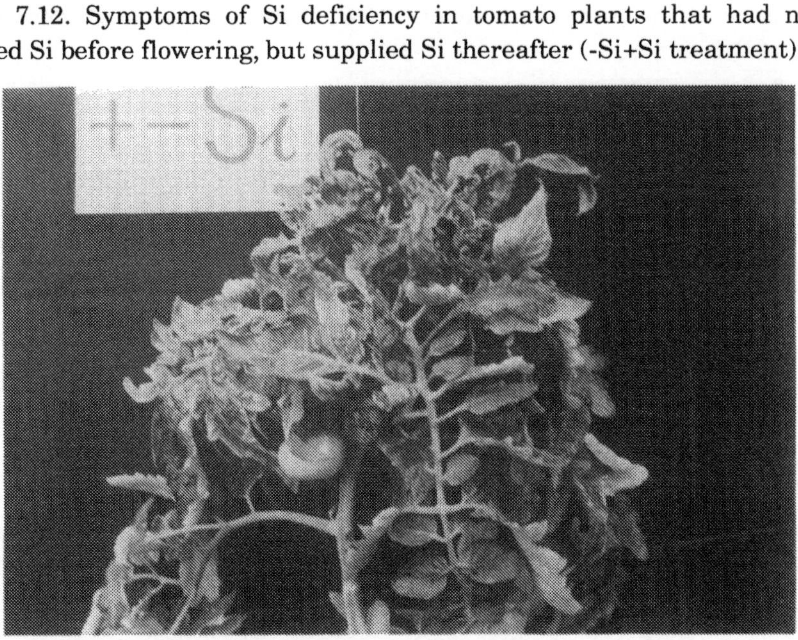

Figure 7.13. Symptoms of Si deficiency in tomato supplied Si before flowering, but not thereafter (+Si-Si treatment).

Table 7.13
Effect of Si supply at various stages on the growth of tomato*

Treatment	Top length (cm)	Root length (cm)	Top weight (g dry wt.)	Root weight (g dry wt.)	Number of fruit	Fruit weight (g fresh wt.)
+Si+Si	108	63	46.2	6.7	4	168
+Si-Si	53	54	37.9	7.4	3	70
-Si+Si	88	59	32.5	3.7	0	0
-Si-Si	45	55	24.3	3.5	0	0

* Tomato variety: Beiju; culture solution : Shive and Robbins No.1 solution 1972-12-24 seeding, 1973-1-15 seedlings were planted in pots and +Si and –Si treatments were started. On Feb.10, 1973, abnormal symptoms appeared in –Si pot, and +Si-Si and –Si+Si treatments were newly set up. Harvested on March 17, 1973.

The Si content of the top without Si supplied was as low as 0.01%, while that with Si supplied was 0.19% Si (Table 7.14). Addition of Si after the flowering stage increased the Si content up to 0.12%. The addition of Si decreased the content of P, K and B.

Figure 7.14. Growth of tomato with various treatments (from left, -Si-Si, -Si+Si, +Si-Si, +Si+Si).

Table 7.14
Effect of Si supply at various time points on the mineral composition of tomato leaves*

	Si %	P %	K %	Ca %	Mg %	Fe ppm	B ppm
+Si+Si	0.19	0.70	2.85	2.74	0.59	103	22
+Si-Si	0.02	0.87	3.17	3.54	0.74	91	12
-Si+Si	0.12	0.76	4.03	2.77	0.55	97	50
-Si-Si	0.01	1.59	5.49	2.85	0.73	100	34

*Culture solution: Shive and Robbins No.1 solution

Tomato is a crop sensitive to environmental conditions. Miyake et al. (1976c) further investigated the effect of sunshine and air temperature and other factors on the expression of the symptoms of Si deficiency in tomato. They found that optimum temperatures with difference for day and night, high light intensity, sufficient aeration, and low planting density are key points for the development of Si deficiency symptoms. When the plant was cultured under high temperature and under insufficient sunshine or at a low temperature and insufficient sunshine, the symptoms of Si deficiency were not observed. The symptoms were also not observed under the conditions without a difference in temperature between day and night. Under the optimum day/night temperatures, the symptoms of Si deficiency enhanced by long-day treatment and reduced by short-day treatment. The symptoms were also observed in plants supplied with Si at 5 ppm SiO_2. The expression of Si-deficiency symptoms varied with the cultivar and cultivation time (Miyake et al., 1976d).

Because the symptoms of Si deficiency observed in tomato resemble those of B deficiency, it was suspected that these symptoms resulted from Si-induced B deficiency. However, even when three times more B was supplied, the same symptoms were observed, suggesting that these symptoms were caused by Si deficiency.

Marschner et al. (1990) argued that the symptoms of Si deficiency in tomato may be caused by Si deficiency-induced Zn deficiency because more P was taken up in the absence of Si and excess P may precipitate with Zn, resulting in decreased availability of internal Zn. However, Miyake et al. (1990) re-examined the Si deficiency symptoms of tomato in the presence of Zn and P at various concentrations. Three concentrations of Zn (0.01, 0.05, 0.10 ppm) and of P (2.3, 0.57, 0.14 mM) were employed (2.3 mM P and 0.01 ppm Zn were the concentrations in the Shive and Robbins solution used in the previous experiments). In the presence of 2.3 mM P and 0.01 ppm Zn in the absence of Si showed abnormal symptoms similar to those reported previously (Miyake and Takahashi, 1978). The growth was markedly improved by the presence of Si. When the Zn concentration was increased to 0.05 and 0.10 ppm, the plants also

Table 7.15
Effects of Zn concentration on the growth of tomato in the presence and absence of Si

Concentration in nutrient solution			Dry weight of leaf	Fresh weight
SiO_2 (ppm)	P (mM)	Zn (ppm)	and stem (g/pot)	of fruit (g/pot)
100	2.3	0.01	42.0	189.3
0	2.3	0.01	29.7	20.8
100	2.3	0.05	41.7	162.1
0	2.3	0.05	38.6	70.6
100	2.3	0.10	44.5	137.6
0	2.3	0.10	39.2	103.2

developed abnormal symptoms in the absence of Si, but the symptoms at 0.10 ppm Zn was not so severe as those of the 0.05 ppm Zn treatment. Dry weight of leaf and stem and fresh weight of fruits were improved as the concentrations of Zn in the nutrient solution increased (Table 7.15).

In the presence of 0.57 and 0.14 mM P the plants supplied with 0,01 ppm Zn developed abnormal symptoms in the absence of Si, but the symptoms at 0.14 mM P were not so severe as those of 0.57 mM P. The dry weight of the leaf and stem and fresh weight of fruits increased as the P concentration in the nutrient solution decreased (Table 7.16). These results suggest that part of the symptoms of Si-deficiency observed in tomato cultured in the solution containing P at a high concentration and Zn at a low concentration might contribute to the Si-deficiency induced Zn deficiency as suggested by Marschner et al. (1990). However, the growth at a low P and high Zn concentration suggest that tomato needs Si for healthy growth.

Table 7.16
Effects of P concentration on the growth of tomato in the presence and absence of Si

Concentration in nutrient solution			Dry weight of leaf	Fresh weight
SiO_2 (ppm)	P (mM)	Zn (ppm)	and stem (g/pot)	of fruit (g/pot)
100	2.3	0.01	41.0	173.9
0	2.3	0.01	30.9	21.3
100	0.57	0.01	41.7	187.5
0	0.57	0.01	35.7	87.3
100	0.14	0.01	44.4	191.4
0	0.14	0.01	31.4	173.1

The symptoms of Si deficiency did not appear during the vegetative stage, but appeared from the first flowering stage, suggesting that Si has some effect on the hormones related to the development and differentiation of plants. The effects of sunshine, day length and temperature on the expression of Si deficiency also support this speculation although the exact mechanisms are unknown.

7.1.4. Cucumber

Cucumber takes up Si passively and the Si content of the leaves increases with increasing Si concentration in the medium. Miyake and Takahashi (1982a, b, c, 1983a,b) investigated the effect of Si on the growth of cucumber. Miyake and Takahashi (1982a, 1983a) grew cucumber in a nutrient solution (as a basal nutrient solution, 1/10 strength of Shive and Robbins solution was used) with (100 ppm SiO_2) or without Si in a growth chamber. Similar to tomato, the plant

Figure 7.15. Initial symptoms of Si deficiency in cucumber (5 weeks after seeding).

without Si supply grew normally like the plants supplied with Si at the earlier growth stage. However, at the flowering stage (about 5 to 6 weeks after sowing), malformation of newly developed leaves (8th or 9th leaf), such as curling, was observed in the plant without Si supply (Figure 7.15). Powdery mildew on the leaves without Si supply was observed (Figure 7.16), while the plant with Si showed a healthy appearance (Figure 7.17).

When the plants showing abnormal symptoms were cultured in a solution with Si (-Si+Si), the newly developed leaves were normal, and there was no powdery mildew damage (Figure 7.18). By contrast, when the plants cultured with Si were transplanted to a solution without Si (+Si-Si), the powdery mildew was observed in the newly developed leaves. Powdery mildew was severe on the plants without Si supply throughout the growth period (-Si-Si), while no such symptoms were observed on the plants supplied Si throughout the growth period (+Si+Si).

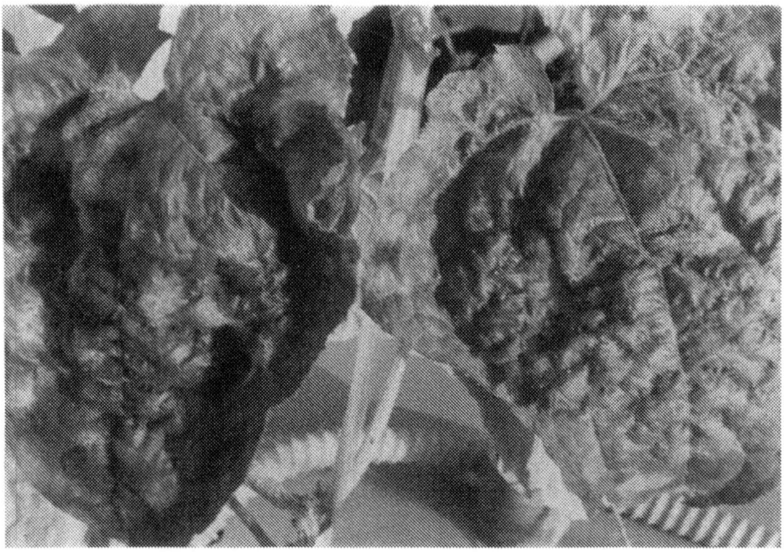

Figure 7.16. Malformed leaves of Si-deficient cucumber (8th and 9th leaves). Many powdery mildew colonies were observed.

Figure 7.17. Abnormal symptoms of Si-deficient cucumber. Left, 100 ppm SiO_2; right, 0 ppm Si.

Figure 7.18. Growth of cucumber with various treatments (from left, -Si-Si, -Si+Si, +Si+Si).

Table 7.17
Effect of Si supply at various stages on the growth of cucumber

Treatment	Top length (cm)	Top weight (g dry wt.)	Root weight (g dry wt.)	Fruit weight (g fresh wt.)	Si content (%) in leaf
+Si+Si	259	46	5.7	466	2.88
+Si-Si	246	43	5.1	397	2.03
-Si+Si	214	32	4.8	321	2.59
-Si-Si	207	33	4.7	370	0.05

Top length, number of leaves, top weight, root weight and fruit weight of the –Si-Si plant were markedly inferior to those of the +Si+Si plants (Table 7.17). These growth parameters of the –Si+Si plants, which were deprived of Si until the flowering stage, were inferior to those of the +Si-Si plant which were supplied with Si until the flowering stage, and were almost the same as those of the –Si-Si plants. The fertility of pollen was significantly reduced by Si-deficiency (Table 7.18).

The effect of Si concentrations in the culture solution on the growth was also investigated. The length and weight of the top and fruit weight increased with increasing Si concentration in the nutrient solution (Table 7.19). The Si content in the leaves also increased from 0.03 to 2.60 % Si with increasing Si concentration from 0 to 100 ppm SiO_2.

The effect of Si on the mineral content was examined in cucumber. The P and K contents of the leaves were markedly decreased by the addition of Si, while those of Ca and Mg were less affected (Table 7.20). Silicon had only a slight effect on the mineral contents of the root.

Table 7.18
Effect of Si deficiency on pollen fertility of cucumber plant

SiO_2 ppm in solution	Time of pollen fertility tested	Fertility ratio (%)
0	just before bloom	85
0	in bloom	83
100	just before bloom	97
100	in bloom	97

Table 7.19
Effect of the concentration of Si in the culture solution on the growth of cucumber plant

SiO_2 ppm in solution	Top length (cm)	Top weight (g dry wt.)	Root weight (g dry wt.)	Fruit weight (g fresh wt.)	Si content (%) in leaves
0	180	64.1	5.6	8	0.03
5	205	70.9	4.5	67	0.21
20	231	78.8	5.0	142	0.64
100	238	94.6	5.4	261	2.60

In addition to the experiments conducted using water culture as described above, the effect of application of Si fertilizer on the growth of cucumber was also investigated at the research farm of Okayama University. Calcium silicate or potassium silicate was applied at 200 or 225 kg per 10a (corresponding to 70 kg/10a soluble Si). As a pH control, equal amount of alkali was applied as

Table 7.20
Effect of Si on the mineral contents of cucumber

	-Si	+Si
P content (%)		
Leaf	2.08	0.86
Stem	0.82	0.84
Root	0.94	1.08
Ca content (%)		
Leaf	5.20	4.31
Stem	1.56	1.60
Root	0.74	1.10
K content (%)		
Leaf	3.36	1.35
Stem	1.12	2.48
Root	3.68	2.24
Mg content (%)		
Leaf	1.68	1.47
Stem	0.94	0.95
Root	0.26	0.38

[a] The plants were cultured in a nutrient solution in the presence or absence of 100 ppm SiO_2 as silicic acid.

Table 7.21
Effect of silicate fertilizers on the growth and SiO_2 content of cucumber expressed as percentages of the control*

Si fertilizers (200 kg/10a)	Harvest Year	Relative value (%)			SiO_2% of leaves
		Dry weight of stems and leaves	Fruit yield	SiO_2 content of leaves	
Calcium silicate	1978	115	118	195	2.88
	1979	125	140	189	2.88
	1980	116	130	204	3.68
Potassium silicate	1978	113	115	182	2.69
	1979	145	145	189	2.86
	1980	116	129	136	2.44

*Control was dressed $CaCO_3$ equivalent to alkalinity of silicate fertilizers

calcium carbonate at 137 kg/10a. The trial was carried out for 3 years successively. Both dry weights of stems and leaves and fruit yield were increased by the application of calcium silicate or potassium silicate compared to the control (calcium carbonate application) each year (Table 7.21). Calcium silicate and potassium silicate showed similar effects on the growth. The Si

Figure 7.19. Initial symptoms of Si deficiency in soybean (cv A62-2, 5 weeks after seeding).

Figure 7.20. Growth of soybean (cv A62-2) with various treatments (from left, -Si-Si, -Si+Si, +Si-Si, and +Si+Si).

content of leaves increased by the application of either calcium silicate or potassium silicate, while there were no large differences in the content of P, K, Ca, and Mg between plants supplied with silicate and carbonate.

Overall, Si has a function in cucumber similar to that in rice in the aspect of increased resistance to disease. On the other hand, unlike that in rice, which benefits from Si from early growth stage, the beneficial effect of Si on the growth in cucumber is expressed after the flowering stage. In this respect, Si behavior in cucumber is similar to that in tomato. Therefore, cucumber is an intermediate type between rice and tomato not only in terms of uptake but also in its effect on the growth.

7.1.5. Soybean

Soybean belongs to Ca-type plant and is able to fix nitrogen by the action of root nodules. Miyake and Takahashi (1985) investigated the effect of Si deficiency on the growth of soybean in relation to the nodulation ability of soybean in a nutrient solution (1/4 strength Shive and Robbins solution). They used a pair of isogenic lines which can or cannot form nodules. Plant not supplied Si showed normal growth at the earlier growth stage as the plant

Table 7.22
Effect of Si on the growth and Si content of soybean at various growth stages

	Line[a]	Treatment[b]			
		+Si+Si	+Si-Si	-Si+Si	-Si-Si
Top length (cm)	A62-1	75	77	74	65
	A62-2	70	62	64	61
Number of pods	A62-1	26	22	26	23
	A62-2	27	25	25	22
Top dry weight (g/plant)	A62-1	28.3	21.5	22.6	21.5
	A62-2	30.6	21.1	20.5	18.9
Root dry weight (g/plant)	A62-1	6.4	5.2	5.7	4.9
	A62-2	7.6	5.4	4.5	4.0
Si content (%)					
Upper leaves	A62-1	0.45	0.08	0.54	0
	A62-2	0.46	0.07	0.48	0
Lower leaves	A62-1	0.71	0.36	0.60	0.005
	A62-2	0.84	0.18	0.67	0
Roots	A62-1	0.11	0	0.09	0
	A62-2	0.06	0.01	0.10	0

[a] A62-1, nodulating line; A62-2, non-nodulating line
[b] At the beginning of the flowering stage, the treatments were subdivided into four, Si newly supplied (-Si+Si), newly removed (+Si-Si), Si continuous supply (+Si+Si), and without Si supply throughout (-Si-Si). Silicon was supplied at 100 ppm SiO_2 as silicic acid.

supplied with Si. However, at the flowering stage, the newly-developed leaves (7th and 8th leaves) showed malformation such as curling and curving to the outside in the Si-deficient plants of both lines (Figure 7.19). When Si was supplied to the plant showing Si deficiency after the flowering stage, the newly developed leaves were normal, while abnormal symptoms appeared in the newly developed leaves when the plant that had been supplied Si was grown in –Si solution (Figure 7.20). These symptoms are similar to those observed in tomato and cucumber as described above.

The growth (top length, number of pods, top weight, and root weight) was increased by the addition of Si (Table 7.22). There was no difference in the effect of Si on the growth between the nodulating cultivar and non-nodulating cultivar. The Si content of the leaves supplied with Si was 0.85% Si, which was much higher than that of tomato, but lower than that of cucumber. Since the Si content of the root is much lower than that of the shoot, soybean seems to take up Si passively like cucumber. The content of P in the leaves was decreased by addition of Si (Table 7.23), while that of K, Mg, N and Ca was hardly affected.

Table 7.23
Effect of Si on mineral content in soybean (cv. A62-1)

	-Si	+Si
P content (%)		
Leaf	2.36	1.15
Stem	0.50	0.44
Root	0.92	1.24
Ca content (%)		
Leaf	3.81	3.17
Stem	0.49	0.50
Root	0.32	0.39
K content (%)		
Leaf	0.92	0.81
Stem	0.57	0.62
Root	1.00	2.47
Mg content (%)		
Leaf	1.60	1.43
Stem	0.25	0.35
Root	1.40	1.45
N content (%)		
Leaf	2.72	2.64
Stem	2.72	3.04
Root	3.58	3.67

The growth parameters increased with increasing Si concentrations from 0 to 100 ppm SiO_2 (Table 7.24, Figure 7.21). No differences were observed in the response to Si between two isogenic lines.

The fertility of pollen from the Si-deficient plant was lower than that from Si-supplied plant in both lines (Table 7.25). The reduction in pollen fertility was higher in the nodulating line than in non-nodulating line.

7.1.6. Strawberry

Miyake and Takahashi (1986) also performed Si-deficiency experiments in strawberry in a nutrient solution (1/10 strength of Shive and Robbins solution). The symptoms of Si deficiency observed in cucumber and tomato were not observed in strawberry, but the growth (dry weight of the shoot and fruit yield) was significantly depressed by Si-deficiency (Table 7.26). The pollen fertility was also reduced by Si-deficiency. The Si content of the leaves of strawberry was similar to that in soybean and higher than that of roots, suggesting that the roots take up Si passively but not rejectively. Similar to tomato, soybean, and cucumber, the content of P in the leaves was remarkably decreased by Si.

Figure 7.21. Growth of soybean (cv A62-2) supplied with various Si concentrations (from left, 0, 5, 20, and 100 ppm SiO_2).

Table 7.24
Effect of Si concentration on the growth of soybean

	Line[a]	Treatment (SiO_2 ppm)			
		0	5	20	100
Top length	A62-1	65	70	75	75
	A62-2	61	66	66	70
Number of pods	A62-1	23	24	27	26
	A62-2	22	22	24	27
Top dry weight (g/plant)	A62-1	21.5	27.8	30.2	28.3
	A62-2	18.9	26.8	29.3	30.6
Root dry weight (g/plant)	A62-1	4.9	7.7	6.5	6.4
	A62-2	4.0	6.2	6.1	7.6
Si content of leaves (%)	A62-1	0.005	0.27	0.38	0.58
	A62-2	0	0.25	0.43	0.69

[a] A62-1, nodulating line; A62-2, non-nodulating line

Table 7.25
Effect of Si deficiency on pollen fertility in soybean

Line	Treatment (SiO_2 ppm)	Fertility ratio (%)
A62-1 (nodulating)	100	97±1
	0	84±3
A62-2 (non-nodulating)	100	96±2
	0	90±2

7.1.7. Bamboos

Bamboo is widely used in Japan. Bamboo culms are used for fence and various bamboo works, and bamboo sprouts are used for cooking.

Ueda et al. (1961) investigated the growth response of bamboo to Si. They found that the genera of bamboos and sasas take up a large amount of Si and most of Si taken up is accumulated on the leaves (Table 7.27). The bamboos (Mousouchiku and Madake) took up a large amount of Si from June to November, and in winter the content of Si in the leaves was constant. The content of Si in the leaves was generally constant regardless of the size and age of the culm.

The content of Si in the leaves in a good grove was higher than that in a poor grove. The soil in a good grove had a large Si-supplying capacity, and Si seemed to have the function to increase the resistance of bamboos to injury by disease (Table 7.28).

Table 7.26
Growth response of strawberry to Si *

		+Si	-Si
Top (g dry weight)		16.0	13.1
Root (g dry weight)		6.5	6.8
Fruits (g fresh wt.)		529	422
Pollen fertility (%)		91	80
Mineral content (%)			
Si	Leaves	0.57	0.03
	Roots	0.03	0.00
P	Leaves	0.37	0.72
	Roots	0.73	0.62
N	Leaves	2.72	3.36
	Roots	1.50	2.11
Ca	Leaves	1.19	0.69
	Roots	0.27	0.21

*+Si treatment was supplied with 50ppm SiO_2 as H_4SiO_4

Table 7.27
Content of SiO_2 in each part of bambo and sasa

	SiO_2 content (%)		
	Leaves	Culm	Rhizome
Phyllostachys edulis (Musouchiku)	5.68~9.00	0.29~0.33	0.25~0.33
Phyllostachys reticulata (Madake)	7.40~10.85	0.47~0.60	0.45~0.55
Bambusa multiplex	8.47~9.57	0.30~0.40	0.60~0.77
Bambusa arundinacea	3.55	1.00	0.45
Sasa paniculata	5.70~8.00	1.50~1.70	0.80~1.03
Shibataea kumasaka (Okamezasa)*	7.40	2.05	0.85
*Pleioblastus pubescens**	15.30	3.00	1.65

*2-year- old; others 3-year-old

The application of calcium silicate fertilizer had a marked effect on the growth of bamboo. The fresh weight of *Shibataea kumasaka* (Okamezasa) was nearly doubled by the application of calcium silicate (Table 7.29).

In a Madake grove (*Phyllostachys reticulata*), the application of calcium silicate increased the number of newly sprouting culms by nearly 40% on the average during 3 years and also increased the hardness of the culm (Table 7.30).

These results indicate that application of calcium silicate fertilizer has a good effect on the production of bamboos.

Table 7.28
Relationship between SiO_2 content in bamboo leaves and the productivity of the grove

	Bamboo species			
	Phyllostachys edulis (Mousouchiku)		*Phyllostachys reticulata* (Madake)	
	SiO_2 (%)	Nos. examined.	SiO_2 (%)	Nos. examined.
Good grove	7.99±1.01	13	10.55±0.97	15
Middle grove	6.07±1.27	23	8.12±1.01	30
Poor grove	5.04±0.63	8	7.09±0.83	13
Injured grove (disease)	5.38±0.63	4	6.00±0.91	11

Table 7.29
Effect of calcium silicate application on the growth of above and under ground parts of *Shibataea kumasaka* (Okamezasa) through two years

	NPK* (control)	NPK+Si
Above the ground		
Number of culms	338	532
Culm length (cm)	20.7	23.9
Fresh weight of top (g) (index)	919 (100)	1813 (197)
Under the ground		
Number of rhizome branches	184	309
Total length (cm)	58.2	106.9
Fresh weight (g) (index)	1350 (100)	2590 (192)

*application of fertilizer: 2 6g N, 15g P_2O_5, 24g K_2O, 75g SiO_2 /plot (0.8 m^2)

Table 7.30
Effect of silicate fertilizer application on the growth* of *Phyllostachys reticulata* (Madake) and hardness of the culm

	NPK* (control)	NPK+Si
Number of new culms (index)		
1st year	39 (100)	61 (156)
2nd year	36 (100)	40 (111)
3rd year	29 (100)	42 (145)
Number of fallen culms		
1st year	12	0
2nd year	8	0
3rd year	0	0
Hardness of the culm (kg/mm^2)		
Yearling culm		
Outside hardness	4.35	4.82
Inside hardness	1.00	1.22
2 years old culm		
Outside hardness	4.37	5.12
Inside hardness	1.20	1.90

*Number of newly sprouting culms in the each plot
Application of fertilizers: 2.2kgN, 0.6kg P_2O_5, 0.6kgK_2O, 2.8kg SiO_2/plot (300 m^2/year)

Table 7.31
Effect of Si on the growth of scouring rush and horsetail. Scouring rush and horsetail were cultured for 226 and 76 days, respectively, with or without Si..

	Scouring rush		Horsetail	
	+Si	-Si	+Si	-Si
Top dry weight (g)	5.6	2.1	3.1	1.9
SiO_2 in the top (%)	8.40	0.15	7.67	0.10

7.1.8. Scouring rush and horsetail

Scouring rush (*Equisetum hiemale* L.) and horsetail (*Equisetum arvense* L.) are known to have high Si contents. Miyake and Takahashi (1976a) investigated the response of these two plants to Si. As shown in Figure 7.22 and Table 7.31, Si -deficiency significantly reduced the growth of both plants. The Si uptake during a period of 24 h by both plants was not affected by transpiration (Table 7.32). When the roots were cut off, the Si uptake was markedly reduced. These results indicate that the response of both scouring rush and horsetail to Si is similar to that of rice.

Sporangia usually form in a natural state, but in this experiment, sporangia were not formed during the cultivation time of scouring rush (226 days) and horsetail (76 days). This may be due to the over supply of nutrients to the culture solution. Therefore, the effect of Si observed in this experiment may be the effect on the vegetative growth.

A

B

Figure 7.22. Effect of Si supply on the growth of scouring rush (*Equisetum hiemale* L., A) and horsetail (*Equisetum arvense* L., B).

Table 7.32
Effect of transpiration and excision of roots on the Si uptake by scouring rush and horsetail

	Treatment		
	Control	Suppression of transpiration	Excision of roots
Scouring rush			
Si uptake[a]	0.41	0.56	0.02
Transpiration[b]	4.60	1.74	0.96
Horsetail			
Si uptake	0.23	0.20	0
Transpiration	7.81	3.10	2.60

[a] mg SiO_2/g root dry wt./24h
[b] ml H_2O/g shoot dry wt/24h

7.2. FUCTIONS OF SILICON

Silicon has not been included in the list of essential elements for higher plants. According to the criteria proposed by Arnon and Stout (1939) for essential elements, a given plant must be unable to complete its life cycle in the absence of the element. However, no evidence has yet been shown that plants are unable to complete its life cycle in the absence of Si. As described above, both Si-accumulating and non-accumulating plants can mature without Si supply although their growth and grain or fruit yield are significantly reduced by Si deficiency. One argument is that Si may function as a micronutrient and that it is not possible to completely remove Si from the growth medium by currently available techniques. However, the fact that the effect is larger when more Si accumulates in the shoots, suggests that quite a large amount of Si is required for the functions of Si. Another criterion for essentiality of elements is that the element must be directly involved in plant metabolism. However, evidence is still lacking on the involvement of Si in the plant metabolism. Nevertheless, Si has a number of functions such as stimulation of photosynthesis, enhancement of tissue strength, and reduction of plant transpiration rate. All these functions contribute to increased dry matter production and resistance of plants to physical, chemical, and biological stresses.

7.2.1. Stimulation of photosynthesis and translocation of photoassimilated CO_2

7.2.1.1. Photosynthesis

Silicon deposited in the leaf blade of rice keeps the leaf erect. Therefore, Si may stimulate canopy photosynthesis by improving light interception. This is particularly important since it helps to minimize mutual shading in dense plant stands and when nitrogen fertilizers are heavily applied,.

Under conditions without mutual shading, the effect of Si on the photosynthesis in rice was investigated by Takahashi et al. (1966b). They used rice which had been supplied with 100 ppm SiO_2 for 0, 5, 10, 15, and 20 days. The experiment for measurement of $^{14}CO_2$ assimilation was performed on a large scale (2000 liter) assimilation chamber at 30°C, light intensity of 50,000 lux, and CO_2 concentration of 800 ppm, ^{14}C 3 mc/4000 L (air). The plants were

Table 7.33
Amount of $^{14}CO_2$ assimilated in plants with various amounts of Si supplied

	Days of Si supply before $^{14}CO_2$ assimilation				
	20	15	10	5	0
Amount of Si in plants (mg)	46.7	34.4	20.3	11.1	-
^{14}C incorporated (10^6 cpm)					
Top					
Et-OH soluble	95.3	85.2	76.7	71.9	68.7
Et-OH insoluble	34.9	34.0	32.3	31.6	29.1
Total	130.2	119.2	109.0	103.5	97.8
Root					
Et-OH soluble	8.8	7.4	7.3	7.2	6.7
Et-OH insoluble	2.7	2.4	2.1	2.1	2.0
Total	11.5	9.8	9.4	9.3	8.7
Whole plant					
Et-OH soluble	104.1	92.6	84.0	79.1	75.4
Et-OH insoluble	37.6	36.4	34.4	33.7	31.1
Total	141.7	129.0	118.4	112.8	106.5
^{14}C cpm/cm^2 leaf area	0.91	0.92	0.91	0.90	0.90

allowed to assimilate for 6 hours from AM 9:00 to PM 3:00. The amount of $^{14}CO_2$ assimilated per individual plant was higher in the plants with a high Si content than those with a low Si content (Table 7.33). However, there was no difference in the amount of $^{14}CO_2$ assimilated per leaf area among plants with a different Si content. The differences in the ratio of ^{14}C in the shoots to that in the roots and ratio of ethanol soluble fraction to ethanol insoluble fraction were not observed.

Ma (1990) measured the photosynthetic rate in the leaf blade of Si-deficient (0.11% SiO_2) and Si-supplied rice (10.1% SiO_2) under natural conditions. There were no detectable difference in the CO_2 assimilation rate between the -Si and +Si plants under any photo flux density although both plants responded to photo flux density up to 1700 mol/m^2/sec (Figure 7.23). The transpiration rate and leaf conductance was also unaffected by Si in rice. Similar results were reported by Kawamitsu et al. (1989). All these results suggest that the effect of Si on the photosynthetic rate is small under optimum growth conditions. However, under water-stress conditions, the photosynthetic rate was reported to be higher in +Si plants than in the -Si plants (Matoh et al., 1991), probably due

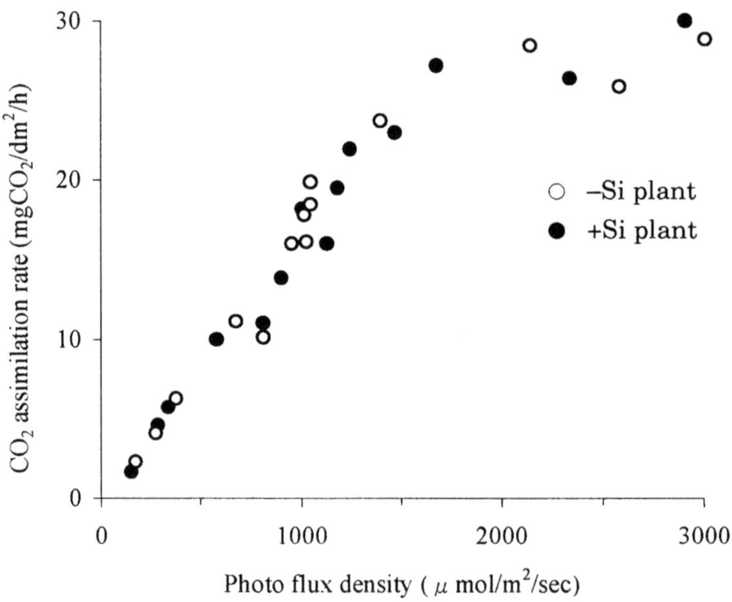

Figure 7.23. Effect of Si on light response curves of CO_2 assimilation rate.

to the Si-induced decrease of transpiration rate.

Silicon was once hypothesized to increase the photosynthetic rate by facilitating the transmission of light (Kaufman et al. 1979). Silica bodies in the leaf epidermal system might act as a "window" to facilitate the transmission of light to photosynthetic mesophyll tissue. However, evidence supporting the "window" hypothesis has not been obtained. Agarie et al. (1996) reported that

Table 7.34
Effect of Si on light absorbance at wavelengths of 400, 500, 600, and 700 nm in rice leaves

SiO_2 conc. (ppm)	SiO_2 content (% of dry weight)	Absorbance (%) at each wavelength			
		400 nm	500 nm	600 nm	700 nm
0	0.29	74.7	93.7	87.4	80.0
20	3.06	70.7	92.4	84.6	77.8
40	5.56	74.0	93.5	87.3	80.1
100	12.60	72.1	92.8	85.8	79.8

Table 7.35
Effect of Si on the light-energy-use efficiency and quantum yield of rice leaves

SiO_2 conc. (ppm)	Light-energy-use efficiency (μmol CO_2/μmolquanta)	Quantum yield (mol CO_2/mol quanta)
0	0.033	0.080
20	0.024	0.058
40	0.025	0.062
100	0.032	0.078

Table 7.36
Effect of Si on the translocation of photoassimilated CO_2 to panicle*

Date of $^{14}CO_2$ Assimilation	Plant parts	^{14}C (cpm×10^6)					
		Ethanol soluble		Ethanol insoluble		Total	
		+Si	-Si	+Si	-Si	+Si	-Si
July 23	Panicle	0.15	0.14	0.31	0.27	0.46	0.41
	Leaf + stem	5.60	4.92	28.03	25.53	33.63	30.45
	Root	0.75	0.77	16.04	16.90	16.79	17.67
	Whole plant	6.50	5.83	44.38	42.70	50.88	48.53
	Ratio of ^{14}C distribution to panicle					0.9%	0.9%
August 31	Panicle	1.23	0.90	5.22	3.94	6.45	4.84
	Leaf + stem	18.05	16.39	79.35	62.38	97.40	78.77
	Root	0.65	0.66	18.51	17.44	19.16	18.10
	Whole plant	19.93	17.95	103.08	83.76	123.01	101.71
	Ratio of ^{14}C distribution to panicle					5.3%	4.7%
September 26	Panicle	61.17	49.56	155.24	104.37	216.41	153.93
	Leaf + stem	362.74	318.71	136.77	129.53	499.51	448.24
	Root	16.89	15.68	8.80	8.49	25.77	24.17
	Whole plant	440.80	383.95	300.81	242.39	741.69	626.34
	Ratio of ^{14}C distribution to panicle					29.2%	24.6%

* ^{14}C in each plant part at the harvest time (September 29). $^{14}CO_2$ was assimilated at the various growth stages of rice plants with or without Si supply.

the optical properties of leaf transmittance, reflectance and absorbance spectra in the Si-supplied leaves were almost equal to those in the non-supplied rice leaves (Table 7.34). Furthermore, the light energy use efficiency and quantum yield of leaves with Si supplied was less than those of leaves without Si supplied (Table 7.35), suggesting that silica bodies do not function as 'windows' in rice leaves.

7.2.1.2. Effect of Si on the translocation of photoassimilated CO_2 to panicle

Takahashi et al. (1966b) investigated the effect of Si on the translocation of photoassimilated CO_2 to panicle. They grew rice in the presence (+Si) or absence (-Si) of Si in an environment-controlled growth chamber and fed $^{14}CO_2$ to the plants at the maximum tiller number stage (July 23), 13 days before (August 31) and after full heading date (September 26). After feeding for 6 hours, the plants were allowed to grow until harvest. ^{14}C assimilated at the maximum tiller number stage was mainly partitioned to the shoot (60%) and the root (30%), but scarcely to the panicle (Table 7.36). By contrast, 5.3 and 29.2% of ^{14}C assimilated before and after the full heading date, respectively, was partitioned to the panicle in the Si-supplied rice, while 4.7 and 24.6% of ^{14}C was partitioned, respectively, in the Si-deficient rice. These results suggest that Si stimulates the translocation of photoassimilated CO_2 to the panicle in rice.

7.2.2. Alleviation of physical stress

Physical stresses include stresses of temperature, light, wind, water, drought, freeze, radiation, ultraviolet ray and so on. A number of studies have shown that Si functions in alleviating these stresses.

7.2.2.1. Radiation injury

Radiant rays cause injury of plant, but the degree of damage is related to physiological status. Takahashi (1966a) investigated the protective effect of Si on radiation injury in rice. Rice seedlings (30 days old, with or without 100 ppm SiO_2) were irradiated with various doses (from 300 to 4800 R) of γ-rays (^{60}Co). After that half of the +Si plants were deprived of Si and half of the –Si plants were supplied with Si and were cultured for another 40 days. The decrease in dry weight was less in the +Si+Si and +Si-Si plants than in the

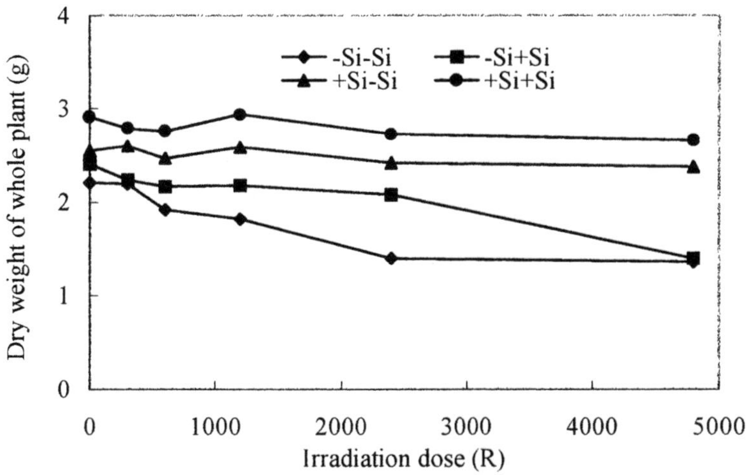

Figure 7.24. Effect of Si application on radiation injury in rice. The plant was supplied with Si before and after radiation treatment.

–Si-Si plants (Figure 7.24), suggesting that Si increases the resistance of rice to radiation stress. Furthermore, when the plants were supplied with Si after radiation treatment (-Si+Si), the growth recovery was faster compared to the plants without Si supply.

7.2.2.2. Water stress

Excess water loss (transpiration) causes stomata closure and therefore decreases the photosynthetic rate. Transpiration from the leaves is made mainly through the stomata and partly through the cuticle. As Si is deposited beneath the cuticle of the leaves forming a Si-cuticle double layer as described before, the transpiration through the cuticle would be decreased by Si deposition. Si can reduce the transpiration rate in rice, which has a thin cuticle (e.g. Okuda and Takahashi, 1961c; Yoshida, 1965; Ma, 1988). Table 7.37 shows the transpiration rate of rice supplied with Si at various concentrations (Okuda and Takahashi, 1961c). The transpiration rate decreased with increasing Si concentrations in the solution. The transpiration rate in rice was negatively correlated with the Si content of the shoot (Table 7.38, Ma, 1988). The effect of Si on the transpiration at various growth stages was also examined

Table 7.37
Effect of Si supply on transpiration rate in rice

Concentration of SiO_2 in culture solution (ppm)	Transpiration rate (g H_2O/g fresh weight·24hr)	
	July 26	September 6
0	4.5 (100)	5.1 (100)
5	3.8 (84)	4.2 (82)
20	3.8 (84)	4.2 (82)
60	3.7 (82)	3.9 (76)
100	3.3 (73)	3.6 (71)
	25th 3pm to 26th 3pm fine weather	5th 11am to 6th 11am fine weather

in rice. The transpiration rate in rice supplied with 100 ppm SiO_2 was 15 and 30 % less during the tillering stage and elongation stage, respectively, compared with rice without Si (Table 7.39, Ma, 1988). All these results indicate that Si reduces the transpiration rate of rice having more than 10% SiO_2 by 20 to 30%.

Table 7.40 shows the effect of Si on rice growth under various humidity (Ma, 1990). Under a water-stressed condition (low humidity), the effect of Si on rice growth was larger than that cultivated under a non-stressed condition (high humidity). The transpiration rate of Si-supplied rice was 30% less than that of Si non-supplied rice at a low humidity.

As mentioned above, among yield components the percentage of ripening is most affected by Si in both rice and barley. This function of Si may be attributed to Si deposited in the hull. One important factor for normal development of spikelets is to keep a high moisture condition within the hull (Seo and Ohta 1982). The Si content in the hull of the rice grain becomes as

Table 7.38
Transpiration rate of rice having various contents of Si*

SiO_2 content in the shoot %	Transpiration (g H_2O/g dry wt.)
0.02	200.3 (100)
1.59	181.7 (90)
10.29	168.0 (84)
13.22	154.4 (77)

*The transpiration rate during 72 h was measured.

Table 7.39
Effect of Si supply on the transpiration rate in rice at various growth stages

	Transpiration rate (g H_2O/g dry weight /7days)	
	Tillering stage	Elongation stage
-Si plants	442(100)	400(100)
+Si plants*	377(85)	280(70)

*Plants were cultured in a nutrient solution containing 100 ppm SiO_2.

high as 7% Si and that of the barley grain is 1.5%. Silicon in the hull also deposits between the epidermal cell wall and the cuticle, forming a cuticle-Si double layer as in the leaf blades. However, differing from leaves, the transpiration occurs only through the cuticle because the hull has no stoma. Silicon is effective in decreasing the transpiration from the hull. Table 7.41 shows the water loss from the excised spikelets at the milky and maturity stages, which were sampled at 10 and 40 days after heading, respectively. The rate of water loss from Si-free spikelets was about 20% higher than that from Si-containing (7% Si) spikelets at each stage. Therefore, Si plays an important role in keeping a high moisture condition within the hull by decreasing the transpiration rate from the hull. This is especially important under water deficiency stress and climate stress as will be discussed below.

Table 7.40
Effect of Si on the growth and transpiration rate of rice under two humidities

Treatment	Shoot dry weight (g/pot)	Transpiration rate (g H_2O/g dry wt)
	Relative humidity at 40%[a]	
+Si	0.91	471.1
-Si	0.73	635.9
+Si/-Si	1.25	0.74
	Relative humidity at 90%[b]	
+Si	4.40	297.6
-Si	4.05	323.3
+Si/-Si	1.09	0.92

[a] grown for 10 days

[b] grown for 30 days

Table 7.41
Effects of Si application on the transpiration from the panicle at various stages*

	Transpiration rate (mg H_2O/g dry weight/1h) at	
	Milky stage	Maturity stage
-Si plants	279 (100)	50.4 (100)
+Si plants**	204 (73)	39.2 (78)

* The excised panicles were placed in an incubator for 1h at 30°C with a relative humidity of 30%
**Plants were cultured in the nutrient solution containing 100 ppm SiO_2

7.2.2.3. Climatic stress

Silicon application in rice is effective in alleviating damage caused by climatic stress such as typhoons, low temperature and insufficient sunshine during the summer season (for a review, see Ma et al. 2001). A typhoon attack usually causes lodging and sterility of rice, resulting in fatal reduction of rice yield. Deposition of Si in rice enhances the strength of the stem by increasing the thickness of the culm wall and the size of the vascular bundle (Table 7.42, Shimoyama, 1958). Also the supply of a large amount of Si increases the breaking strength of culm especially when a large amount of nitrogen is supplied (Table 7.43, Iwata and Baba, 1962). Thus Si is effective in preventing lodging.

Strong winds also cause excess water loss from the spikelets, resulting in sterility. Silicon deposited in the hull is effective in preventing excess water loss. In addition, the effect of Si on the rice yield is also obvious under low temperatures and insufficient sunshine stress. As shown in Table 7.44, the

Table 7.42
Effects of Si on the thickness of culm wall and the size of vascular bundle

	Treatment		Treatment	
Culm wall thickness (μm)	N	728	3N	638
	NSi	781	3NSi	622
	N2Si	826	3N2Si	659
Vascular bundle size ($100\mu m^2$)	N	202	3N	179
	NSi	239	3NSi	171
	N2Si	282	3N2Si	183

Note: N: ammonium sulfate 5 g/pot (3N: 15g/pot), Si: sodium silicate 10 g/pot (2Si: 20 g/pot), Rice cv: Aichi-asahi

Table 7.43
Effect of Si supply on the breaking strength of culm at the ripening stage

Internode* (downward)	Breaking strength (kg)			
	20 ppm N		40 ppm N	
	50 ppm SiO_2 (A)	200 ppm SiO_2 (B)	50 ppm SiO_2 (C)	200 ppm SiO_2 (D)
1	1.00	1.08	0.83	1.09
2	0.88	0.95	0.84	0.92
3	0.90	0.98	0.86	1.03
4	1.14	1.26	0.99	1.23

*Including leaf sheath

effect of Si on rice growth (fresh weight) under shaded conditions is larger than that without shading (Ma et al., 2001), but the mechanisms responsible for this phenomenon are unknown.

7.2.3. Improvement of resistance to chemical stress

Chemical stresses include deficiency and excess of nutrients, low and high soil pH, metal toxicity, pesticide, herbicide, and so on. Silicon has been reported to have important functions in improving the resistance of plants to these chemical stresses.

7.2.3.1. Nutrient-imbalance stress

7.2.3.1.1. Excessive N stress

Application of nitrogen fertilizers is an important practice for increasing rice yield. Silicon has been reported to raise the optimal level of nitrogen in rice. For example, field trials conducted in Hiroshima, Okayama and Shikoku Agricultural Experiment Stations, showed that the optimal level of nitrogen

Table 7.44
Effect of Si on the fresh weight of rice shoot under shading (light-interception coefficient of 52%) and that under a non-shaded condition for 20 days.

Treatment	Shoot fresh weight (g/pot)	
	No shading	Shaded
-Si	3.76	1.45
+Si	4.14	2.03
+Si/-Si	1.10	1.40

Table 7.45

Effect of silicate fertilizer application on the growth of rice plants supplied with nitrogen fertilizer at various rates

Nitrogen application rate (kg/ha)	Stem weight (kg/plot)		Paddy weight (kg/plot)		Number of white head per hill		SiO_2 content of leaves (%)	
	+Si	-Si	+Si	-Si	+Si	-Si	+Si	-Si
0	11.80	10.05	8.20	6.86	1	2	24.5	16.3
40	15.05	12.62	9.60	8.08	2	2	22.3	15.1
60	15.69	15.26	9.69	7.54	4	3	20.8	14.8
80	17.18	15.39	9.99	8.19	1	14	20.5	14.5
100	18.64	17.69	10.60	7.69	6	16	20.1	14.0
120	19.82	18.53	10.16	7.47	8	41	19.5	12.6
140	20.77	19.66	11.07	5.39	18	102	18.8	10.0

+Si plots were supplied with calcium silicate fertilizer equivalent to 2 t/ha.

was 75 kg N per hectare when Si fertilizer was not applied, but this level was raised to 150 kg N per hectare when Si fertilizer (slag) was applied. Table 7.45 shows the effect of Si on rice yield in the field fertilized with nitrogen at various rates at the experimental farm of Kyoto University (Okuda and Kawasaki, 1958). When Si fertilizer was not applied, the yield was the highest at 80 kg N per hectare, and was reduced by application of 140 kg N per hectare. However, when Si fertilizer was applied, the yield increased as the rate of N application was increased to 140 kg N per hectare.

Table 7.46

Effect of Si supply on the growth of rice at various N levels*

Dry weight (g/hill)	20 ppm N			40 ppm N		
	50 ppm SiO_2 (A)	200 ppm SiO_2 (B)	B/Ax100	50 ppm SiO_2 (C)	200 ppm SiO_2 (D)	D/Cx100
Top	88.87	94.68	106.5	106.08	118.46	111.7
Leaf blade	14.66	16.53	112.8	21.53	26.36	122.4
Leaf sheath and stem	53.50	56.00	104.7	60.27	65.87	109.3
Ear	20.71	22.51	108.7	24.28	26.23	108.0
Root	10.12	9.19	90.8	9.68	9.37	96.8
Total	98.99	103.87	104.9	115.76	127.83	110.4

*Harvested at the ripening stage

Iwata and Baba (1962) also investigated the effect of Si on the adaptability of rice for heavy nitrogen application with special reference to photosynthesis. Experiment was carried out using aggregate plants (plant density 25 cmx25 cm) which were solution-cultured under four combinations of nitrogen and Si levels. A high Si supply level (200 ppm Si plot) increased the dry weight of the top, and the effect was greater at the 40 ppm N plot than 20 ppm N plot (Table 7.46).

Excessive application of nitrogen makes the leaf blades droopy, resulting in mutual shading and thereby reduction of photosynthesis. As shown in Figure 7.25 (Takahashi, 1982) and Table 7.47 (Yoshida et al., 1969), the angle between the culm and the tip of the leaf increased as the amount of nitrogen applied was increased, but decreased with the increase of Si applied.

Excessive nitrogen also increases susceptibility to diseases and lodging. Silicon deposited on the stems and leaf blades is effective in preventing lodging, diseases and mutual shading as stated above. Thus, Si endows rice with the adaptability for heavy nitrogen application.

A

B

C

Figure 7.25. Growth of rice supplied nitrogen at various levels with or without Si (100 ppm SiO_2). A, 5 ppm N; B, 20 ppm N; and C, 120 ppm N.

Table 7.47
Relationship between Si and N supply and leaf erectness in rice plants (cv. IR8) at the flowering stage

N supply ($mg\ L^{-1}$)	Si supply ($mg\ SiO_2\ L^{-1}$ as sodium silicate)		
	0	40	200
		Angle	
5	23°	16°	11°
20	53°	40°	19°
200	77°	69°	22°

*Angle between the culm and the tip of the leaf

The effect of Si on the quality (low protein content) of rice was also investigated recently. Miyamori (1996) reported that application of silica gel resulted in decreased protein content in rice grain by increasing production efficiency of brown rice of nitrogen. He estimated that half paddy fields of Hokkaido are deficient in available Si for production of rice with low protein (below 80 g/kg).

7.2.3.1.2. Deficiency of P and excess stress

The beneficial effects of Si under P-deficiency stress have been observed in many plants including rice and barley. Early observations from a long-term field experiment conducted at Rothamsted Experimental Station showed that the barley yield was higher in a field applied sodium silicate than in a field without Si application when P fertilizers were not supplied. In an experiment with nutrient solution, Si supply increased the dry weight of rice shoot more at a low P level than at an intermediate P level (Okuda and Takahashi, 1962c; Ma and Takahashi 1990a).

Such beneficial effects of Si might be attributed to several factors including Si-improvement of P availability in soil and in plant, and increased P uptake by Si. Silicon may affect the P availability of soil by displacing fixed P and/or reducing P fixation by masking active Al and Fe. However, in previous studies using sodium or calcium silicate as a Si source, which application caused pH increase in soil, while increased pH may also affect soil P availability. To discriminate the effect of Si and that of pH on the P availability in soil, a comparative study was conducted by applying both silicic acid and sodium silicate as Si sources to a P-deficient soil (available phosphate, 6.0 mg P_2O_5/100

Table 7.48
Dry weight and mineral content of rice shoot grown in a P-deficient soil with or without silicic acid supplied under flooded or upland condition

	Upland		Flooded	
	-Si	+Si	-Si	+Si
Dry weight (g/pot)	1.34	1.52	1.40	1.61
Mineral content				
Si (%)	1.44	2.81	2.10	3.58
P (ppm)	974	880	1041	1030
Fe (ppm)	96.1	84.6	126	116
Mn (ppm)	1870	1518	3696	2761
P/Fe	10.1	10.4	8.3	8.9
P/Mn	0.52	0.58	0.28	0.37

g soil) (Ma and Takahashi, 1990b, 1991). Addition of silicic acid to the soil did not cause change of soil pH. When rice was grown in this soil under either flooded or upland conditions, the growth (shoot dry weight) was increased by the addition of Si (Table 7.48). However, addition of Si did not increase the P content of the shoot. The Mn content was significantly decreased by Si, resulting in a high P/Mn ratio in the plant, while the Fe content was relatively unchanged. In terms of P availability of soil, previous application of silicic acid at various concentrations did not affect the P adsorption by the soil (Table 7.49). Neither the P displacement by Si was increased by increasing Si concentrations in the soil with or without P supplied (Figure 7.26). These results suggest that Si as silicic acid does not affect the P availability of soil and P uptake by the roots and that the beneficial effects of Si on the growth result from improved availability of internal P in the plant by decreasing Mn uptake.

Table 7.49
Effect of a previous application of silicic acid on P adsorption

Treatment[a]	P adsorbed (μmol g^{-1} soil)
Si0	175.2
Si1	173.9
Si2	175.2
Si3	175.5

[a] Si0, Si1, Si2 and Si3 represent soil samples which had previously received 0, 0.23, 0.47 or 0.94 mg Si g^{-1} soil as silicic acid.

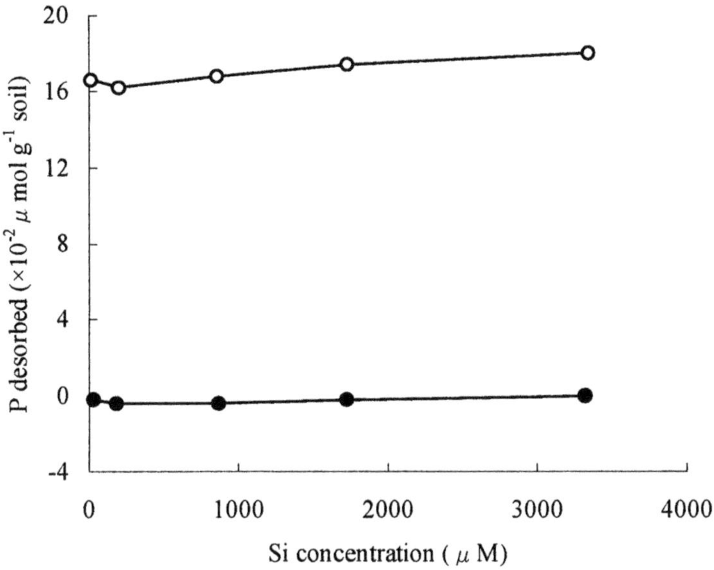

Figure 7.26. Effect of Si at various concentrations on P displacement in a P-deficient soil which had previously been received 0 (●) or 4 mg P/g soil (○).

When sodium silicate or sodium carbonate was applied to the P-deficient soil, the pH of soil was raised by 1 unit. The growth of rice on the soil was increased by either sodium silicate or carbonate applications, but the increase by sodium silicate was greater than that by sodium carbonate under both flooded and upland conditions (Table 7.50). Neither the application of sodium silicate nor that of sodium carbonate increased the P content in the shoot under either condition. Application of sodium silicate significantly decreased the Mn content, resulting in a higher P/Mn ratio in the plant. By contrast, application of sodium carbonate did not affect the Mn concentration. Application of either sodium silicate or carbonate increased the N content nearly 2 times. Neither the P adsorption by the soil nor P displacement by Si was affected by sodium silicate and carbonate (Figure 7.27, Table 7.51). Similar to the application of Si as silicic acid, these results also suggest that Si as silicate does not affect the P availability of soil and P uptake by the roots, and that the beneficial effects of Si is attributed to improved availability of internal P in the plant by decreasing Mn uptake. The mechanism responsible for decreased Mn uptake will be

Table 7.50
Dry weight and mineral content of rice shoot grown in a P-deficient soil with or without calcium silicate or carbonate supplied under flooded or upland condition

	Upland			Flooded		
	None	Sodium carbonate	Sodium silicate	None	Sodium carbonate	Sodium silicate
Dry weight (g/pot)	1.08	1.15	1.29	1.05	1.32	1.53
Mineral content						
Si (%)	1.57	1.24	2.62	1.98	1.61	2.88
P (ppm)	1141	1064	1088	1250	1236	1365
N (%)	2.41	3.97	4.00	2.09	4.35	4.37
Fe (ppm)	62.4	66.7	63.9	89.2	98.9	91.9
Mn (ppm)	2152	2314	1473	3715	3836	2949
P/Fe	18.3	16.0	17.0	14.0	12.5	14.9
P/Mn	0.53	0.46	0.74	0.34	0.32	0.46

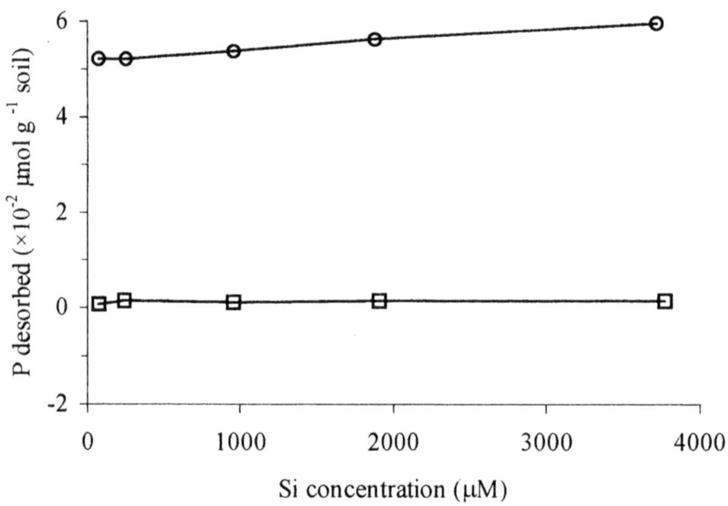

Figure 7.27. Effect of Si (as silicate) concentration on P desorption in a P-deficient soil which had previously received 0 (□) or 4 mg P g^{-1} soil (○). The P desorption was measured by equilibrating 10 g soil for 5 days at 20°C with 25 ml of 0.1 M NaCl containing a range of Si concentrations as indicated in the Figure.

Table 7.51
Phosphorus adsorption by a P-deficient soil with or without previous application of sodium carbonate and sodium silicate.

Treatment	P adsorbed[a] (μmol g^{-1} soil)
None	171.1
Sodium carbonate	172.1
Sodium silicate	169.3

[a] Determined by equilibrating 2 g soil with 30 ml of 0.01 M $CaCl_2$ containing 12 mM P for 5 days at 20°C.

described later. In addition, a part of the beneficial effect of the application of sodium carbonate is attributable to the pH effect, which stimulates the ammonification of organic nitrogen in soil. Silicon is present in the form of silicic acid in soil solution, which does not dissociate at a pH lower than 8. Therefore, it is unlikely that interaction between silicic acid and phosphate (anionic form) occurs in the soil.

Table 7.52
Effect of Si supply on the growth and grain yield of rice grown in a nutrient solution containing various levels of P.

Dry weight (g)	P_2O_5 in the nutrient solution (ppm)		
	2	10	50
Panicle			
+Si	4.7	7.5	7.2
- Si	2.9	6.5	5.7
Leaf+stem			
+Si	8.5	11.7	12.8
- Si	7.8	11.1	11.9
Root			
+Si	3.8	4.3	4.2
- Si	4.1	4.4	4.5
Whole plant			
+Si	17.0	23.5	24.2
- Si	14.8	22.0	22.1
Ripened grain			
+Si	3.6	6.0	5.9
- Si	1.7	5.2	4.3

Table 7.53
Effect of Si supply on the P content of rice grown in a nutrient solution containing P at various levels

P_2O_5 in the nutrient solution (ppm)	P content (%)							
	Panicle		Leaf+stem		Root		Whole plant	
	+Si	-Si	+Si	-Si	+Si	-Si	+Si	-Si
2	0.09	0.12	0.02	0.02	0.02	0.02	0.03	0.03
10	0.30	0.38	0.11	0.17	0.19	0.27	0.18	0.24
50	0.50	0.64	0.41	0.74	0.64	0.77	0.48	0.73

Okuda and Takahashi (1962c) investigated the effect of Si on rice growth and yield under a low (2 ppm P_2O_5), medium (10 ppm), and high (50 ppm) level of P. They found that the effect of Si on rice growth and yield was more obvious under a low P level than that under a medium level (Table 7.52). There was no difference in the P content of the shoot between the plant with and that without Si application under a low P level (Table 7.53). Ma and Takahashi (1990a) confirmed this result in rice during short-term growth (1 month) (Figure 7.28). They also found that Si decreased both Fe and Mn uptake, resulting in a higher P/Fe ratio and especially higher P/Mn ratio in the plant under a low P level (Table 7.54). The availability of P is controlled by the levels of Mn and Fe in plants when the P concentration is low. Phosphorus is translocated and redistributed in plants as inorganic P. Since P has a high affinity with metals such as Fe and Mn, the, internal availability of P could be controlled by the level of Mn, Fe and other metals when the concentration of P is low. The results from soil culture and water culture indicate that the larger beneficial effect of Si on plant growth under P-deficiency stress may be attributed to the enhanced availability of internal P by decreasing excess Fe and Mn uptake. This is supported by the fact that Si supply increased the rate of P translocation to the panicles in rice (Table 7.55, Okuda and Takahashi, 1962c). When rice was exposed to radioactive ^{32}P for 6 days at the panicle forming stage, the total uptake of ^{32}P and its translocation to the top did not differ among treatments, but distribution of ^{32}P into the panicle was much higher in the plant supplied with Si above 60 ppm.

Excess P stress hardly occurs in natural soils, but in some house soils where P fertilizers are heavily applied or in nutrient solution culture where P is supplied at a high concentration. When rice was given P at a high

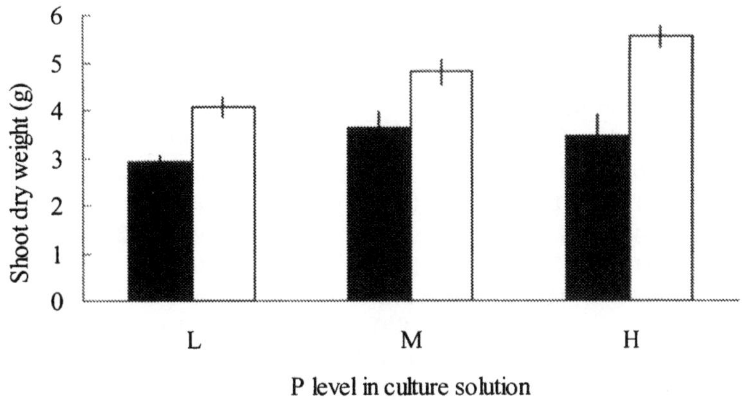

Figure 7.28. Dry weight of shoots grown in a nutrient solution with (□) or without (■) Si at three P levels. L, M and H represent Low (0.014mM), Medium (0.21mM) and High (0.70mM) P levels, respectively. Vertical lines on columns represent SD.

concentration, chlorosis was observed on the leaves (Figure 7.29), probably due to decreased availability of essential metals such as Fe, Zn. However, in the presence of Si, the chlorosis did not occur at a high P concentration. This beneficial effect of Si might be attributed to the lower P uptake caused by Si. When Si was supplied to rice at a high P level, the P content of the shoot was markedly decreased (Okuda and Takahashi, 1962c, Table 7.53). Ma and Takahashi (1990a) further found that the content of organic-P in rice was not affected by Si, but that of inorganic-P was significantly decreased by Si in the presence of a high level of P (Figure 7.30). Deposition of silicon in the roots and/or Si-induced decrease of transpiration rate may be responsible for the decreased uptake of P in the medium containing P at a high concentration.

Table 7.54
The P/Fe and P/Mn ratios in rice shoots grown on a nutrient solution containing P at various concentrations with or without Si (100 ppm SiO_2).

P level (mM)	P/Fe		P/Mn	
	-Si	+Si	-Si	+Si
0.014	12	17	3	9
0.21	59	57	13	19
0.70	113	80	23	29

Table 7.55
Effect of Si on the distribution of ^{32}P in rice grown in a nutrient solution containing 0.09 mM P*

SiO_2 (ppm) concentration in solution	Counts of ^{32}P (10^3 cpm) and distribution (%) in each part					Panicle/(Leaf+stem) (%)
	Panicle	Leaf+stem	Top	Root	Total	
0	28.6	126.2	154.8	95.3	250.1	22.6
	(11)	(51)	(62)	(38)	(100)	
5	36.4	135.6	172.0	86.8	258.6	26.8
	(14)	(52)	(66)	(34)	(100)	
20	38.3	137.4	175.7	83.9	259.6	27.9
	(15)	(53)	(68)	(32)	(100)	
60	52.9	104.4	157.3	93.6	250.9	50.7
	(21)	(42)	(63)	(37)	(100)	
100	72.0	92.8	164.8	91.9	256.7	77.7
	(28)	(36)	(64)	(36)	(100)	

* Plants were fed ^{32}P at the ear-forming stage for 6 days

Figure 7.29. Alleviative effect of Si on the toxicity of excessive P. Rice was grown in a nutrient solution containing 6.5, 50, and 200 ppm P_2O_5 in the presence (100 ppm SiO_2) or absence of Si.

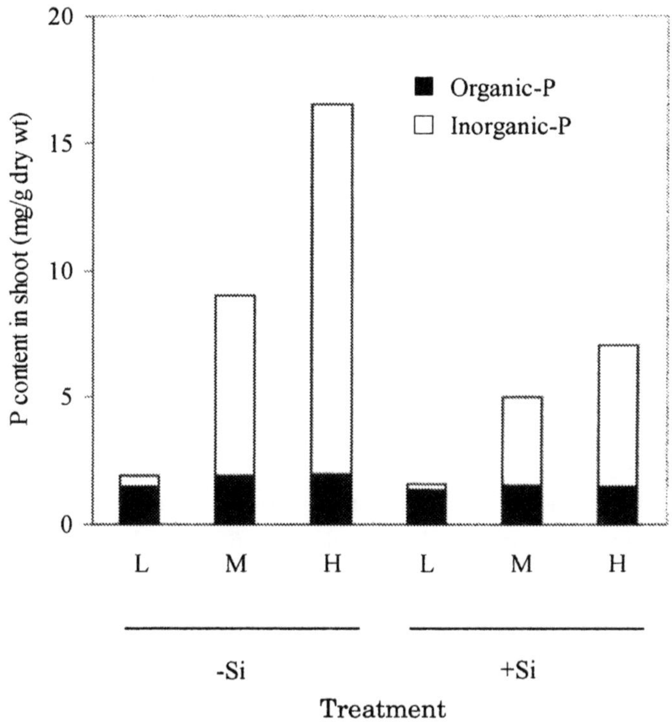

Figure 7.30. Content of organic-P and inorganic-P in the shoots of rice plants supplied 0.014 (L), 0.21 (M), and 0.7mM P (H) in the presence and absence of Si (1.67mM Si as silicic acid).

The Si-induced decrease of P uptake is also observed in some Si non-accumulating plants including tomato, soybean, strawberry and cucumber as described above, although the mechanisms are unknown.

7.2.3.2. Metal toxicity

7.2.3.2.1. Excess Na

Since Si reduces the transpiration rate in rice by 20-30% as described above, there is a possibility that Si suppresses the translocation of salt from the rhizosphere to the shoot and thereby alleviate salt stress. Matoh et al. (1986) examined this possibility in rice. They grew rice in a nutrient solution with or

Table 7.56
Dry matter production and the contents of Na and Cl in rice grown under 100 mM NaCl salinity with and without Si addition

Treatment		Plant parts	Dry matter (g)	Na (ppm)	Cl (ppm)
-Si	Control	Shoots	2.69	1720	3370
		Roots	0.52	1940	1080
	+100mM NaCl	Shoots	1.16	35100	55900
		Roots	0.22	13400	11600
+Si	Control	Shoots	2.97	1460	3410
		Roots	0.44	1590	1190
	+100mM NaCl	Shoots	1.93	19000	43700
		Roots	0.30	16700	17900

without Si (100 ppm SiO_2) in the presence of 100 mM NaCl for one month. The salt-induced reduction in the growth was 60% in the plant without Si, but 35% in the plant with Si (Table 7.56). The concentration of Na in the shoot was decreased to about half by Si addition, suggesting that Si suppresses the translocation of Na from the root to the shoot. Tsuda et al. (2000) reported that the low Si deposition in the spikelets was responsible for the occurrence of white heads under salinity conditions in rice.

7.2.3.2.2. Fe Toxicity

In the experiment of Okuda and Takahashi (1961b, c), the deficiency of Si significantly increased the contents of Fe in the shoot of rice, but did not cause the abnormal symptoms reported by Wagner (1940). Wagner used 60 ppm Fe in the nutrient solution, which was much higher than that usually used in rice culture (2 ppm). Therefore, there is a possibility that the abnormal symptoms observed by Wagner resulted from excessive Fe due to Si deficiency. To examine this possibility, Okuda and Takahashi (1962a) investigated the effect of Si on the growth of rice at excessive concentrations of heavy metals. They found that Si added as silicic acid at 100 ppm SiO_2 did not alleviate Cu and Co toxicity, but alleviated Fe and Mn toxicity.

As shown in Figure 7.31A, symptoms of Fe toxicity appeared at 60 ppm Fe in the solution without Si supply and this symptoms was similar to that reported by Wagner. However, addition of Si significantly reduced Fe uptake and

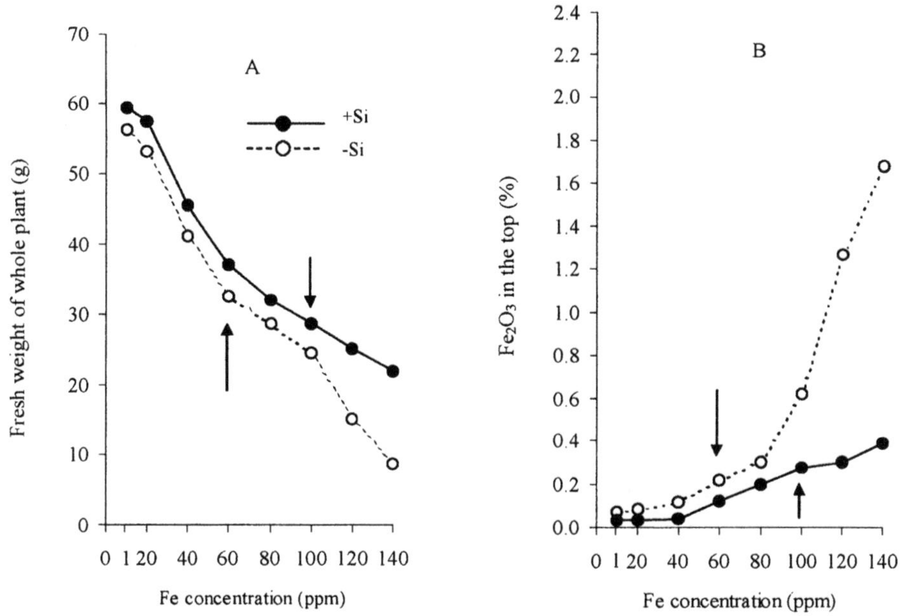

Figure 7.31. Effect of silicon on the alleviation of ferrous-ion toxicity and the iron uptake by lowland-rice. ↑ ↓ shows the iron concentration in solution when symptom of toxicity appeared.

improved the growth (Figure 7.31B). Okuda and Takahashi (1962b) investigated the mechanism of the reduction of Fe uptake by Si. They examined the uptake and distribution of ^{59}Fe, oxidative capacity of the root cell sap for ferrous Fe, oxidative capacity of excised roots for α-naphtylamine, Fe uptake from ferrous Fe and production of ferric Fe in the solution to which the intact plant or excised top was exposed, in the rice plants with various Si contents. The presence of Si during the Fe uptake period significantly reduced ^{59}Fe uptake from ferrous Fe (Figure 7.32). Furthermore, in the absence of Si during the Fe uptake period, less ^{59}Fe was taken up in the Si-containing plants and the ^{59}Fe uptake decreased with increasing Si content (from 0.2 to 7.0% SiO_2) of the shoot. Si had no effect on the oxidative capacity of the root cell sap for ferrous Fe, or the oxidative capacity of excised roots for α-naphtylamine, but the production of ferric Fe in the solution to which the intact plant was exposed was higher in the Si-containing plant (Figure 7.33). However, no difference in the Fe uptake was observed between the plants with Si and those without Si when

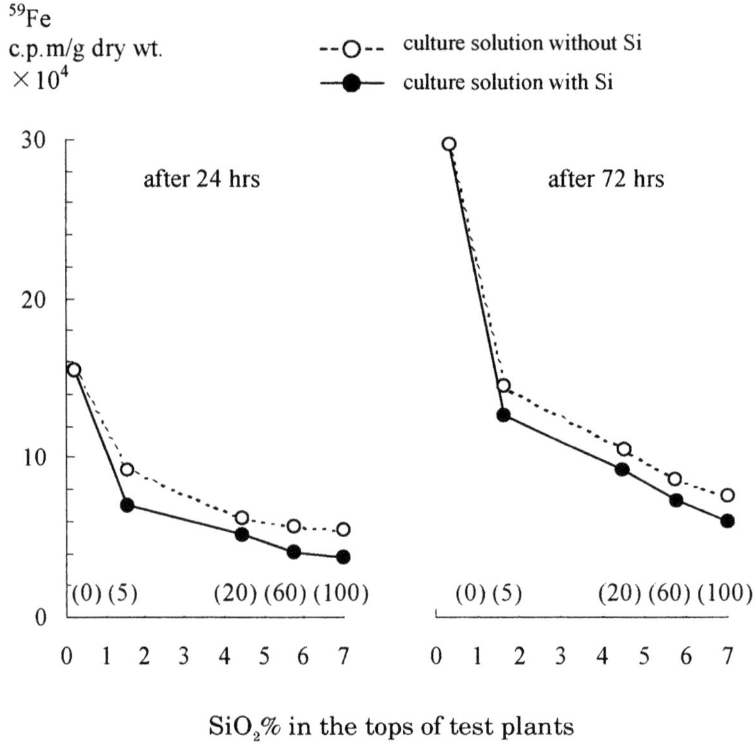

Figure 7.32. Effect of the absorbed Si on the uptake of ^{59}Fe by rice plants from Fe^{2+} solution. ( ) shows SiO_2 concentration (ppm) in the culture solution.

the excised tops were exposed to the ferrous Fe solution. These results suggest that Si enhances the oxidizing capacity of the roots probably by promoting oxygen supply from the shoot to the root, resulting in oxidation of ferrous Fe to ferric Fe on the root surface, thereby suppressing excess uptake of Fe and translocation of Fe from the root to the shoot.

7.2.3.2.3. Mn Toxicity

Silicon has been shown to alleviate the toxicity of Mn in water-cultured rice (Okuda and Takahashi 1962a), barley (Williams and Vlamis 1957; Horiguchi and Morita 1987), bean (Horst and Marschner 1978), and pumpkin (Iwasaki and Matsumura 1999). Three mechanisms seem to be involved depending on the plant species. In rice, similar to the toxicity of Fe, the toxicity of Mn on the

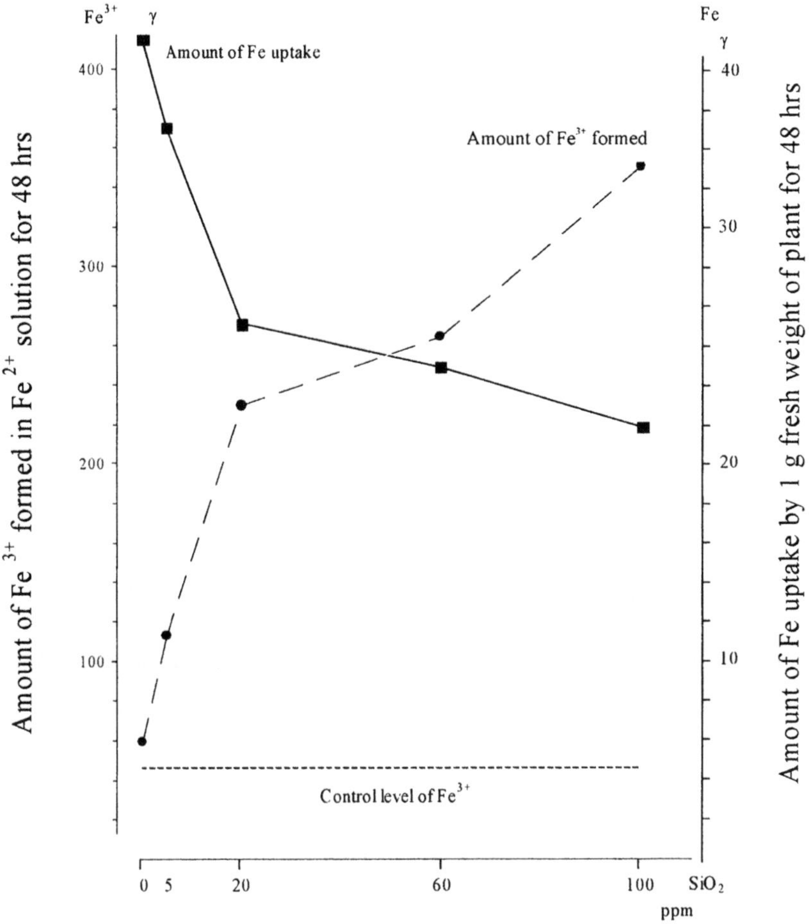

Figure 7.33. Effects of the absorbed SiO_2 on the amount of Fe taken up by the rice plants and the rate of Fe^{3+} formation from Fe^{2+} in the solution to which the plants were exposed.

growth was alleviated and Mn uptake was reduced by Si in the solution containing Mn at a high concentration (Figure 7.34). The reduction of Mn uptake by Si may be caused by the promoted Mn oxidizing capacity of the roots as described above (Okuda and Takahashi 1962a). In barley (Horiguchi and Morita 1987), Si did not reduce Mn uptake, but caused homogenous

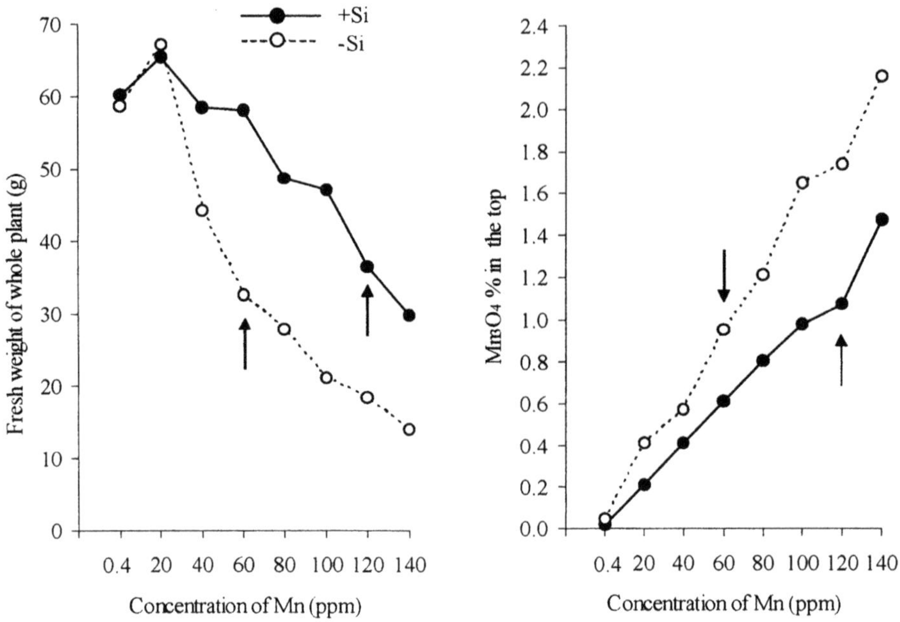

Figure 7.34. Effect of silicon on the alleviation of manganese toxicity and the manganese uptake by lowland rice. ↑ ↑ shows the manganese concentration in the solution when symptoms of toxicity appear.

distribution of Mn in the leaf blade, although the mechanism for this homogenous distribution is still unknown. By contrast, the third mechanism was postulated by Iwasaki and Matsumura (1999) based on the results of the experiments with pumpkin.

Cucumber in Japan has been produced by grafting onto pumpkin stocks. Grafting on the bloom-type stocks that produce a white powder of silica (blooms) on the fruit surface has been replaced by grafting on the bloomless-type stocks, because cucumber fruits with blooms are not preferred by consumers in Japan. However, the cucumbers produced using the bloomless-type stocks showed an increased occurrence of Mn toxicity that has been attributed to rejective uptake of Si by the pumpkin stock (Yamanaka and Sakata, 1993, 1994). Iwasaki and Matsumura (1999) conducted a series of experiments to investigate the mechanism of Si-induced alleviation of Mn toxicity in two cultivars of pumpkins.

Table 7.57
Growth and the contents of Si and Mn in two pumpkin cultivars

	-Si				+Si			
Mn conc. (μM)	10		125		10		125	
	Shoot	Root	Shoot	Root	Shoot	Root	Shoot	Root
Dry weight (g/pot)								
Shintosa	26.8	4.0	10.8	1.4	31.3	4.2	29.3	4.8
Superunryu	29.6	3.1	10.4	1.0	30.9	2.5	19.7	1.8
Si content (mg/plant)								
Shintosa	23.6	1.3	10.1	1.1	167.6	6.0	163.4	5.8
Superunryu	38.9	1.1	16.8	0.8	45.8	8.9	24.7	6.1
Mn content (mg/plant)								
Shintosa	4.5	0.6	19.0	2.1	4.6	0.5	27.6	3.7
Superunryu	4.9	0.4	19.2	2.9	4.3	0.3	25.7	2.1

Table 7.57 shows the effects of Si and Mn on the growth, and the uptake of Si and Mn in bloom-type (cv. Shintosa) and bloomless-type (cv. Superunryu) cultivars of pumpkin stocks. Exposure to high Mn significantly inhibited the growth of the root and shoot of both cultivars in the absence of Si. However, in the presence of Si (1.67 mM as silicic acid), high Mn did not inhibit the growth of Shintosa, but decreased the shoot dry weight of Superunryu by 40%. Silicon did not affect Mn uptake of either cultivar, but Shintosa took up a larger amount of Si than Superunryu.

Electron probe X-ray microanalysis of the leaf of Shintosa grown without Si showed that Mn was accumulated around the necrotic lesion between veins and around the base of the trichomes. However, in the presence of Si, both Mn and Si were accumulated only at the base of trichomes and the accumulation of Mn was confined to that region more distinctly than that in the absence of Si. These results indicate that alleviative effect of Si on Mn toxicity in Shintosa is brought by a high ability to accumulate Si and translocate it to the shoots. This in turn may cause localized accumulation of Mn together with Si in a metabolically inactive form at the base of trichomes.

7.2.3.2.4. Al toxicity

Aluminum (Al) toxicity is a major factor limiting crop production in acid soils. Ionic Al inhibits root growth and nutrients uptake. In an experiment with maize, Si addition as silicic acid significantly alleviated Al-induced inhibition of

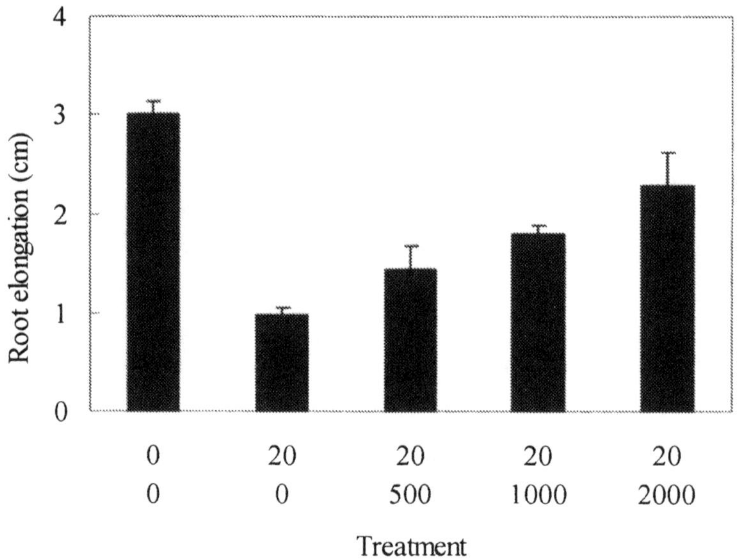

Figure 7.35. Effect of Si on Al-induced inhibition of root elongation in maize. Roots were exposed to 20 μM $AlCl_3$ for 20 h in the presence of silicic acid at various concentrations.

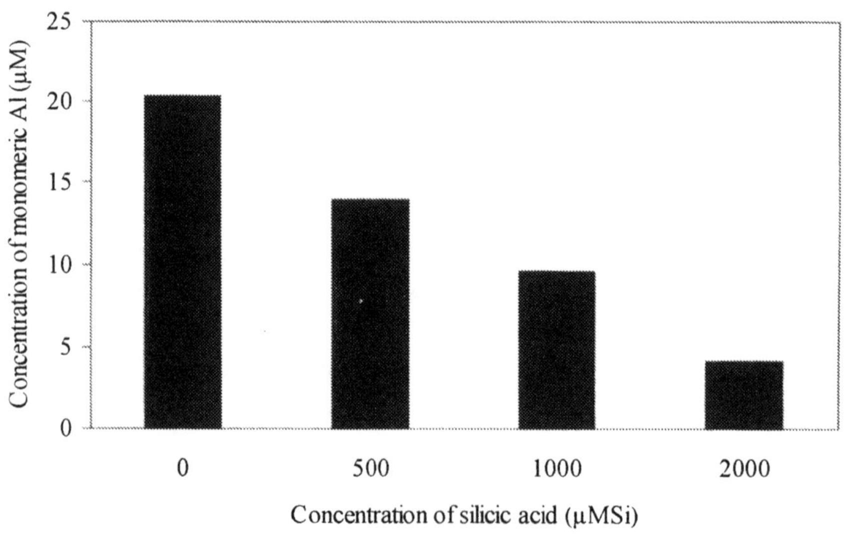

Figure 7.36. Effect of silicic acid on the concentration of toxic Al^{3+}.

root elongation (Ma et al., 1997, Figure 7.35). The alleviative effect was more apparent with increasing Si concentration. The concentration of toxic Al^{3+} is decreased by the presence of silicic acid (Figure 7.36). These results suggest that interaction between Si and Al occurs in solution, probably by formation of Al-Si complexes, a non-toxic from. In rice, the alleviative effect of Si on Al was also reported (Gu et al., 1999).

7.2.4. Increase of resistance to biotic stress

The effect of Si on the resistance to rice blast, brown spot and so on has been observed in the very early period. A few examples on the protective effect of Si on diseases and pests are introduced here.

7.2.4.1. Disease

Yoshii et al. (1958) found that the damage from rice stem rot caused by *Leptosphaeria salvinii* Cat. was alleviated by Si supplied (Table 7.58). Internal soluble N was low and the ratio of total C to soluble N was high in Si-supplied rice. They considered that Si reduced the damage from the stem rot was related to these changes in soluble N.

Rice blast caused by the fungus *Magnaporthe grisea* (Hebert) Barr is one of the most serious threats to rice production. The alleviative effect of Si on blast has been reviewed by Ishiguro (2001). The blast occurrence is usually serious in a year with cold summer. Table 7.59 shows the results of field experiment conducted in 1981, which had a cold summer in North-east Japan (Ohyama, 1985). Silicon was supplied as compost (36 ton per hectare) and calcium

Table 7.58
Effect of Si supply on the severity of rice stem rot caused by *Leptosphaeria salvinii* Cat.

SiO_2 in the irrigation water (ppm)*	Severity of rice stem rot	Soluble nitrogen (sN) and carbohydrate (C) contents (dry weight basis) in the rice shoot		
		sN %	C %	C/sN ratio
0	138	0.52	34.2	66
250	106	0.40	39.7	100
500	64	0.20	33.1	164

*sodium silicate solution was adjusted to pH 5.8 by dilute HCl

Table 7.59
Effect of Si and N application on blast disease in rice

N fertilizer applied (kg/10a)	Grade of infection by blast disease		Leaf blade SiO_2 %**		Leaf blade N%**	
	-Si	+Si*	-Si	+Si	-Si	+Si
0	6.4	2.6	6.5	9.4	2.30	2.26
3.6	9.5	1.7	4.5	9.2	2.38	2.14
7.2	16.7	2.6	3.9	7.9	2.40	2.39
10.8	19.3	5.0	3.3	7.8	2.73	2.24

* calcium silicate application 180kg/10a.
**content at full heading time.

silicate (1.8 ton per hectare). With increasing the amount of nitrogen fertilizer, the severity of leaf blast and ear blast increased (Table 7.59). Application of compost and silicate fertilizer resulted in significant increase in the Si content of the leaf blade and great reduction in the severity of both leaf and ear blast although the N content of leaf blade was only slightly decreased by Si

Table 7.60
Effect of Si on the resistance of cucumber plant to powdery mildew disease

		Treatments (Si ppm)				
		0	5	20	50	100
Si content in leave (%)		0.04	0.05	0.31	1.02	1.50
Leaf-position	1	3+	2+	+	+	±
	2	4+	2+	2+	+	±
	3	3+	4+	2+	+	±
	4	+	3+	2+	±	-
	5	2+	2+	+	±	-
	6	+	+	±	±	-
	7	±	±	±	±	-
	8	-	-	-	±	-
	9	-	-	-	-	-
upward	10	-	-	-	-	-

Disease symptom: number of colonies/leaf area (cm^2), +, <0.1; 2+, 0.2-0.5; 3+, 0.6-1; 4+, >1

application. There was a negative correlation between Si and N content. Application of calcium silicate caused an increase in the ratio of Si to N, resulting in suppression of blast. In a cold summer year, rice has a high N content but low Si content because Si uptake is reduced at low temperature. This is why Si application in a cold summer year is more effective to reduce blast damage.

Silicon has also been reported to prevent powdery mildew and stem rot diseases of cucumber. Miyake and Takahashi (1982c, 1983b) examined the effect of Si at various concentrations on the resistance of the cucumber to powdery mildew. Colonies of powdery mildew appeared in the leaves of the plants supplied with 0 and 5 ppm SiO_2 for 17 days after seeding and developed vigorously thereafter. However, in the plants supplied with 50 and 100 ppm SiO_2, very few or no colonies appeared (Table 7.60, Figure 7.37). The Si content of leaves was negatively correlated with the severity of powdery mildew, suggesting that high Si concentration is required to suppress powdery mildew. According to their results, the Si content in the leave must be higher than 1.5% Si to prevent powdery mildew.

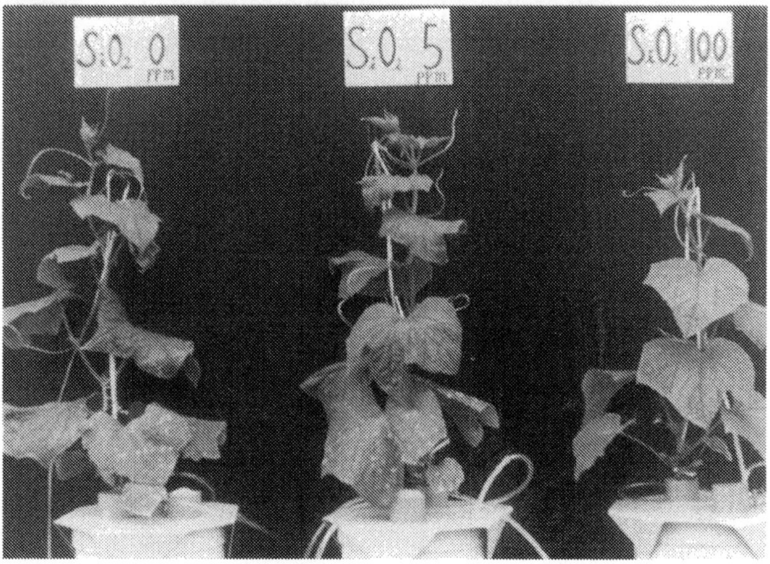

Figure 7.37. Effect of Si on the resistance of cucumber to powdery mildew. Left and center (0 and 5 ppm SiO_2), many powdery mildew colonies were observed. Right (100 ppm SiO_2), colonies were hardly observed.

Table 7.61
Effect of silicate fertilizers on the incidence of Tsuruware disease in cucumber

Treatment	Incidence in	
	1978	1979
Control*	37 %	20 %
Calcium silicate**	20 %	0
Potassium silicate**	11 %	0

* $CaCO_3$ equivalent amount to alkalinity of silicate fertilizers
** 200 kg fertilizer/10a

In the field, the effect of application of Si fertilizer on the suppression of fungal diseases was also observed. At a later growth stage (August), Tsuruware disease (wilt disease caused by *Fusarium* infection) broke out every year in the control, but the disease was significantly suppressed by the application of either calcium silicate or potassium silicate (Table 7.61).

Silicon deposited on the tissue surface as described in the previous chapter may be responsible for the protective effect of Si on biotic stress. It functions as a physical barrier and prevents physical penetration and/or makes the plant cell less susceptible to enzymatic degradation by fungal pathogens.

7.2.4.2. Pest

Silicon suppresses insect pests such as the stem borer, brown planthopper, rice green leafhopper, and whitebacked planthopper, and non-insect pests such as leaf spider and mites. Sasamoto (1958, 1960, 1961) investigated the relationship between Si content of rice and behavior of stem borer (*Chillo suppressalis* Walker). He found that the stems attacked by the stem borer contained less Si. The large jaws of the stem borer gnawing rice with a high Si content would wear out more easily than that gnawing rice with a low Si content. These protective functions of Si may be attributed to Si deposited on the tissue surface.

He further investigated the behavior of the borer in a Petri dish containing rice stem pieces with various Si contents, which was prepared by application of silica gel. He found that most larvae moved to the stem with low Si content. The Si content of the rice stem was negatively correlated with the number of larvae bored into the stem and the amount of feces (Table 7.62). When the stem was extracted with water, the number of larvae that moved to the extract from the stem was smaller in the stem containing a higher Si. The stem

Table 7.62
Effect of silica supply on the resistance of rice plants to the rice stem borer *Chillo suppressalis* Walker

	Amounts of silica gel supplied (g/pot)			
	0	1.5	4.5	6.0
SiO_2 % in the stem	1.35	1.71	2.02	2.11
Number of the larvae bored into the rice stems	22	7	4	2
Amounts of feces (mg)	139	29	11	9

*40 of the fourth instar larvae were incubated in each Petri dish containing 5 cut stems of various SiO_2 contents. Number of the larvae bored into the rice stems were counted after 24 h of inoculation

supplied heavily with N contained some compounds that were extractable with alcohol or ether and attracted the borer, and the biosynthesis of such compounds seems to be suppressed by Si application.

7.3. WORKING PROCESS OF BENEFICIAL EFFECTS OF SILICON ON PLANT GROWTH

The functions of Si in higher plants are summarized in Figure 7.38. Most of these functions are based on the research on rice. It is clear that most functions of Si are mediated by Si deposited on the leaves, stems and hulls. Therefore, differing from the essential elements, the functions of Si in plants are mechanical and not physiological.

As stated above, the effects of Si are more obvious under stress conditions. Plants are exposed to various biotic and abiotic stresses in the field during the growth period. Therefore, Si in plants certainly plays an important role in maintaining healthy growth. It should be kept in mind that an absence of an apparent response to Si fertilizers does not mean that Si does not function in plant growth. This is because Si is abundant in soil and the requirement of Si for healthy growth of most plant species is supplied from the soil.

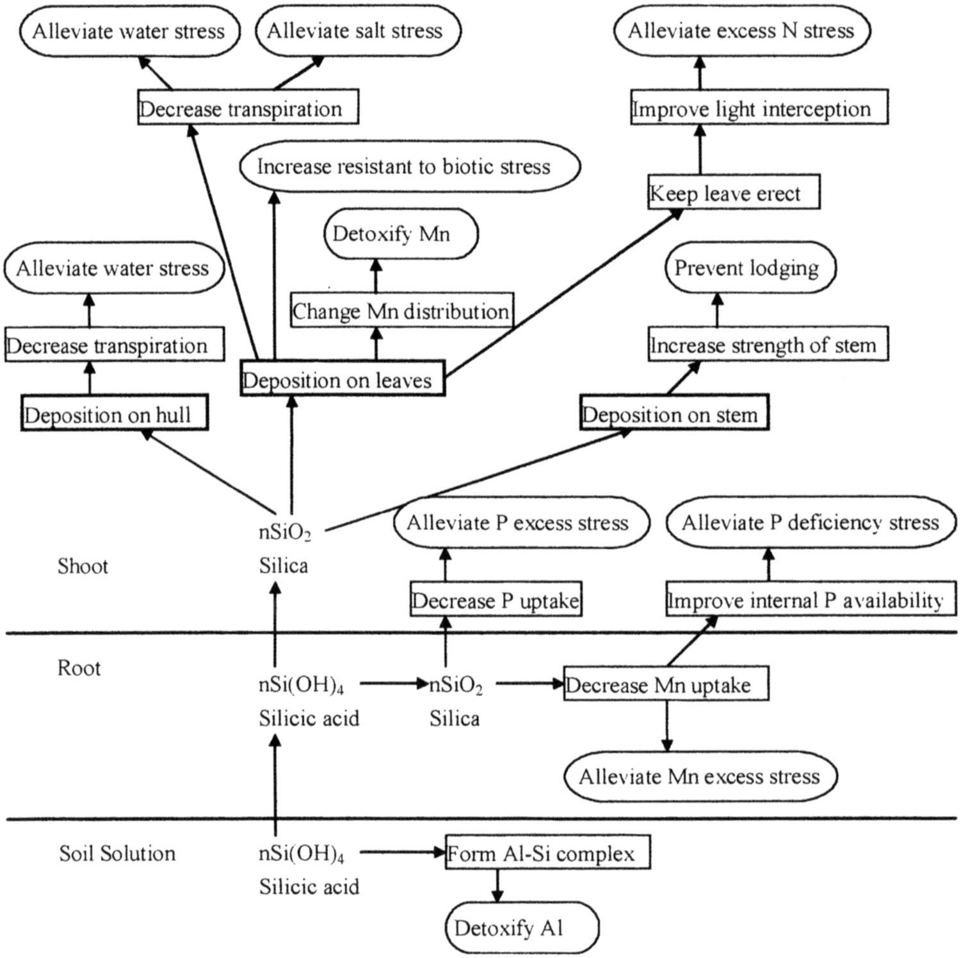

Figure 7.38. Beneficial effects of Si on plant growth in relation to various stresses.

Chapter 8

Summary and prospect of silicon research

As described in the previous chapters, Japanese researchers have made great contributions to Si research, especially in rice. In this chapter, the major achievements in Si research in Japan are summarized and the prospects for future research are described.

8.1. MAJOR ACHIEVEMENTS AND PROSPECT OF RESEARCH ON SILICON IN SOIL

Silicon is abundant in soil, but its solubility is not high enough to supply Si to rice, which requires a large amount of Si for healthy growth. Therefore, soils in quite large areas of rice production in Japan are deficient in Si. Furthermore, rice in Japan is cultivated with heavy application of nitrogen fertilizer and at a high density, where a large amount of Si is especially required for high production of rice. Based on these facts, Si is regarded as an "agronomically essential element" for rice in Japan and Si is applied as a fertilizer in paddy soils.

8.1.1. Survey on Si fertility

In 1955, the Ministry of Agriculture, Forestry and Fisheries of Japan conducted a nationwide survey on the Si content of flag leaves of rice sampled from 37,949 places. The Si content of the flag leaf varied widely, ranging from lower than 7% SiO_2 (about 5% of total samples) to higher than 23% SiO_2 (about 9% of total samples). Such a variation in Si content of flag leaves is related to the variation in the Si-supplying capacity of paddy soils. There is a trend that the Si supplying capacity of paddy soils is higher in north-east regions of Honshu (the Main Island) and Kyushu, and low in middle to west regions of Honshu. About that time, Kobayashi investigated the water quality of major 380 rivers and found that the Si concentration in rivers is high in Hokkaido, north-east regions of Honshu and Kyushu, which corresponds to the Si content of the flag leaf.

A part of the Si in soil is eluviated by percolating water and then moved to rivers. In paddy fields, a large amount of irrigation water (14,000 tons per ha) from the rivers is required. Silicon concentration in irrigation water is therefore reflected in the Si availability in basin geology and greatly contributes to Si fertility in paddy soils. From a pot experiment, soils derived from volcanic ash and shale were found to be rich in soluble Si, and soils derived from aged volcanic ash, quartz porphyry and granite, and peat are poor in soluble Si. These geological factors are well reflected in the regional variation in the Si content in the flag leaf and in the Si concentration in river water.

Application of compost has been considered to increase the Si fertility of soil. According to the results from a long-term field experiment conducted at Shiga Prefectural Agricultural Experiment Station, successive application of compost for 40 years at a rate of 20 tons per ha per year, resulted in 21% and 42% increase of Si uptake by rice and wheat, respectively, and also 24% increase in the content of soluble Si in the soil. These results indicate that successive application of compost at a high rate is important to maintain Si fertility of soils.

Imaizumi and Yoshida (1958) proposed a criterion for the application of silicate fertilizers. According to this criterion, the application of silicate fertilizers is expected to be effective if the Si content of rice straw at harvest in the field is less than 13% SiO_2. The data from a nationwide survey revealed that soils need Si supplement account for one third of paddy soils in Japan. After this survey, the application of inorganic silicate fertilizers was initiated.

Half a century has passed since the first application of silicate fertilizer. During this period, soil Si fertility has been gradually changed with decreased application of compost and silicate fertilizers. The rate of compost application which was 6 ton per ha 50 years ago, has been decreased to one third of this amount. The consumption amount of silicate fertilizer has been gradually decreased to less than one fourth of the peak which was 1.3 million tons in 1968. Therefore, Si fertility of soil is being feared nowadays.

In fact, it was recently found the content of soluble Si in soil and the amount of Si in rice straw have decreased in Yamagata prefecture, one of the major areas for rice production. The Si concentration in irrigation water investigated in 1996 was only half to one third of that in 1956 in all regions within the prefecture. Since Si is essential for stable and high production of rice, a new nationwide survey is required for the Si concentration in irrigation water and Si content of rice straw in order to understand the changes of soil Si fertility.

8.1.2. Method for evaluation of available Si in paddy soil

For evaluation of Si available to rice in paddy soil, the 1 N acetate-buffer (pH 4.0) method established by Imaizumi and Yoshida (1958) had been adopted for a long time. However, with the wide spread application of calcium silicate fertilizers, it was found that this method did not necessarily reflect the Si uptake by rice. This method was originally aimed to extract Si contained in the silicate mineral in soil, while this extraction condition is too strong for Si contained in calcium silicate applied in soils..

Several new methods for estimation of available Si in soil have been developed. One of them is "incubation under a submerged condition" which was developed by Takahashi (1986). This method was aimed to estimate the Si solubilized from soil under flooded condition. As this method is better correlated with the Si taken up by rice, it was adopted in a book titled "Standard Method of Soil Analysis". However, since the acetate-buffer method has been used in many studies until now, a further comparison of these two methods is necessary.

Silicon uptake by rice is affected by many factors and the process of Si solubilization in soils is also complicated. Therefore, it is very difficult to have an all-around method for extraction of soil Si available to rice. For practical use, a convenient and rapid method is needed, while the "incubation under the submerged condition method" takes one week for incubation. Although the "phosphate buffer method" was proposed as a rapid method recently, a better extraction method is needed. In addition, since the amount of Si solubilized from soil varies with the soil, a local method rather than a universal method might be necessary for precise diagnosis of soil Si fertility.

8.2. MAJOR ACHIEVEMENTS AND PROSPECTS OF RESEARCH ON SILCON FERTILIZER

8.2.1. Utilization of slag as a silicate fertilizer

Degraded paddy fields are distributed widely in Japan and the supply of silicate has been reported to improve the soil on these fields. From many trials, Ohta (1953) found that slag, which is derived from the iron and steel industry, can be utilized as a Si source. As a result, in 1955, "silicate fertilizer" was authorized by the Minister of Agriculture, Forestry and Fisheries and the standards for this fertilizer were established. Slag, with a major component of calcium silicate, has been used as a liming material in Europe and America. In Japan, however, slag is used as a silicate fertilizer.

The amount of slag consumed increased rapidly from 1962 when the application of compost as a major Si sources started to decrease. In 1968, the amount of slag consumed reached a peak of 1.3 million tons per year although it gradually decreasing thereafter. Application of slag plays a certain role in covering decreased supply of Si from the compost.

It should be noticed that the slag contains Ca at a higher rate than Si and also contains other metals although their contents are low. These side-components play more active physiological roles than Si. Since Si is applied at a high rate (usually 1 to 2 tons per hectare), the effect of these side-components also should be taken into account. The effect of side-components in slag is positive in some cases, but negative in other cases.

8.2.2. Development of new silicate fertilizers

Slag used as a silicate fertilizer is the waste of iron industry, and the content of Si in the slag is guaranteed. In addition to slag, several new silicate fertilizers have been developed. Fused magnesium phosphate fertilizer has been manufactured in Japan since 1950 as a non-sulfate phosphate fertilizer, to replace superphosphate. It contains 16-26% of citrate soluble Si and that having higher than 20% of citrate-soluble Si was recognized as a silicate fertilizer.

Potassium silicate has been manufactured as a slow-release fertilizer of potassium since 1978. The official standard of commercial fertilizer requires that the fertilizer contain higher than 25% of 0.5 M HCl-soluble Si. Different from slag, the application rate of Si is controlled by the application rate of P or K in these fertilizers.

Silica gel appeared in 1999 as a pure Si fertilizer without containing any side components. Application of silica gel in nursery bed is effective for raising healthy rice seedlings. However, because the price of silica gel is higher than other silicate fertilizers, its application range is limited.

The standard application rate of slag is 1 ton per ha, which costs 25,000 yen. The average yield of rice is 5 ton per ha, and if the percentage of increase in rice yield by the application of slag is 5%, it corresponds to 250 kg per ha. The producer's rice price is 250 yen for 1 kg rice, therefore, the gross income will increase 62,500 yen per ha by application of slag. The benefit of slag application is 2.5 times more than the cost. However, as the application rate of slag is large, a labor-saving method of application needs to be established.

8.2.3. Evaluation of rice straw as a Si source

If rice straw and husk are returned to the soil, most of the Si taken up by the rice from soil may be supplemented. The problem is whether the Si in rice straw and husk is available to the rice plant. A short-term experiment showed that only 10% of the total Si in rice straw could be released, while long-term trial indicated that successive application of compost is effective to increase soil Si fertility. The relationship between Si availability in rice straw and husk and the degree of rotting and solubilization process of Si in soil need to be examined.

8.3. MAJOR ACHIEVEMENTS AND PROSPECTS OF RESEARCH ON SILCON IN PLANTS

8.3.1. Distribution of Si-accumulator in plant kingdom

Many species in Gramineae contain a large amount of Si and the term Si-accumulator has also been used for a long time. However, information is lacking about the distribution of Si-accumulators in the plant kingdom. With the cooperation of several botanic gardens, Takahashi's group analyzed the mineral content of a large number of plants grown under similar soil conditions. The results showed a characteristic distribution of Si accumulators in the plant kingdom.

To discriminate Si-accumulators from other species, Takahashi's group used the Si content of 0.5% as the first criterion and the Si/Ca ratio of 1 as the second criterion. The Si content of 0.5% was used as for the following reason. If the average Si concentration in the soil solution is 10 ppm and average water requirement of plant is 500 ml, the content of Si taken up by passive transport will be 0.5% on a dry weight basis. The Si content varies greatly with the plant species, and the species containing higher than 1.0% Si (two times of the critical value) is termed as Si-accumulator, while those having less than 0.5% Si as Si-excluder. And plant species which the Si content is between 0.5 and 1.0%, is termed as intermediate type. In *Angiospermae*, Si-accumulators containing a large amount of Si, had much lower contents of Ca and B than the other elements. The ratio of Si to Ca content or the ratio of Si to B content was clearly higher in Si-accumulators than in the others. Since the data for the B content of all samples were not available, the Si/Ca ratio was used as a criterion of Si accumulators. Si-accumulators should not only have Si content higher than 1%, but also Si/Ca ratio higher than 1. If the Si content is higher than 1%, but the Ca content is higher than Si content (Si/Ca ratio <1), this plant does not belong to Si-accumulator, but to intermediate type. And if the Si content is lower than 0.5%, but the Ca content is lower than Si content (Si/Ca>1), this plant does not belong to Si-excluder, but to intermediate type.

The data on 400 species revealed that, the distribution of Si accumulators corresponded well to the class, order, and family in the plant kingdom. For example, *Lycopsida* and *Equisetopsida* in *Pteridophyta* are all Si accumulators. In *Filicopsida*, both Si accumulators and non-accumulators are included, but there is a clear separation at the family level. All species in *Gramineae* belong to Si accumulators, but the degree of Si accumulation differs among subfamilies. The order of Si accumulation degree is: *Bambusoideae* >*Pooideae* >*Panicoideae* >*Eragrostoideae*. In *Cyperacea*, which is close to *Gramineae*, most of them are Si-accumulator, but some species are Si-excluders. Although only two species of *Bryophyta* were tested, both showed a high degree of Si accumulation. However, as the contents of Al and Fe were also high in these samples, the possibility of contamination from the soil could not be excluded. Therefore, the degree of Si accumulation in *Bryophyta* remains to be examined further.

The plants growing on nearly the same type of soil were used for measurement of mineral content. Therefore, the difference in the content of minerals among plant species can be attributed to the characteristic uptake by each plant species. It is also well known that some plants accumulate non-essential elements such as Na, Al, and Se in their tops. Accumulation of Al, Na, and Se is the result of adaptation in special soil environments (saline soil, acid soil, and Se-rich soil). By contrast, Si is always abundant in soil. Therefore, Si accumulation depends on the Si-absorbing ability of the plant species. In the course of evolution, plant groups having a high Si absorbing ability might preserve this ability and make it beneficial for survival. This might be the reason why the distribution of Si accumulators fitted well to the phylogenetic tree.

A Si-accumulator is characterized by a low content of Ca and B in addition to a high content of Si. Similarly to Si, both Ca and B are localized at the cell wall, but the relationship between Si and Ca or B is not well understood.

In addition to the Si-accumulators examined by Takahashi's group, other Si-accumulators have also been reported. Therefore, we need more data from a larger number of plant species to discuss the distribution of Si accumulators in the plant kingdom. Since the studies by Takahashi's group were mostly done using the leaves of herbaceous plants, the Si content in woody plants also needs to be investigated. Furthermore, as Si content usually increases with age, it is also necessary to consider the sampling time and leaf position.

8.3.2. Form of Si taken up by rice plants and the mechanism of uptake

In soil solutions with a pH lower than 8, Si is present in the form of uncharged molecule, silicic acid (H_4SiO_4). Takahashi et al. compared the uptake of Si from silicic acid and silicate anion and found that the uptake from silicic acid is significantly superior to that from silicate anion. Furthermore, the uptake of silicate anion is affected by phosphate and nitrate but that of silicic acid does not, suggesting that the form of silicic acid is advantageous for Si uptake by rice.

There are three uptake modes for Si, that is, active, passive, and rejective uptake; they are observed in rice, barley, and tomato, respectively (Okuda and Takahashi, 1965). The uptake mode is determined by the action of the roots. When intact plants are grown in a nutrient solution containing Si, uptake by rice results in a significant decrease in Si concentration in the nutrient solution, while uptake by tomato results in increase in the Si concentration. Uptake by barley does not cause any change in the Si concentration in the nutrient solution. However, when the roots were cut off, the Si uptake by the excised shoot of all species showed passive uptake. The evidence that the Si concentration in the bleeding sap of rice is much higher than that in the external solution also supports the view that rice roots have an active uptake system for silicic acid. By contrast, in tomato, a long time is required for Si in the bleeding sap to reach the same level as that in the external solution.

The Si uptake by rice is affected by metabolic inhibitors. Some inhibitors such as NaCN, DNP, 2,4-D, and iodoacetate inhibit the uptake of both Si and P, while some inhibitors including Na-malonate and Phloridin affect P uptake but not Si uptake. These facts indicate that the system for uptake of Si is different from that for P although both are energy-dependent active uptake processes. The strong inhibition of Si uptake as well as P and K uptake by H_2S also supports energy-dependent uptake of Si in rice.

Radioisotopes are powerful tools to investigate the uptake mechanism of a mineral. Mitsui et al. attempted to produce a radioisotope of Si in a nuclear reactor and utilize it for uptake studies. However, because the half-life is too short (2.6 h), the utilization of this radioisotope is difficult. This problem was later overcome by using a radioisotope of Ge. Ge is an analogue of Si and has a radioisotope with a long half-life. A series of studies revealed that plant roots can not discriminate Si from Ge in terms of uptake although Ge is toxic to the plant growth.

Recently, the specific uptake of Si by rice is further characterized by Ma's group. The Si uptake seems to be mediated by a transporter. From kinetic studies, the K_m for Si uptake is estimated to be 0.32 mM, suggesting the participation of a low affinity transporter. Lateral roots play an important role in the uptake of Si, but root hairs do not play any demonstrable role. A mutant of rice defective in Si uptake has also been isolated and characterized.

Although there are three different modes for Si uptake depending on plant species, the mechanisms responsible for these different uptake modes are unknown. Isolation and identification of genes responsible for Si uptake in rice may be very interesting topics for the near future. A rice mutant defective in Si uptake would provide a powerful tool for identification of these genes. Also, the mechanisms responsible for rejective uptake should be addressed in future in relation to the properties of cell wall and cell membrane. Interestingly, a genotypical difference in Si uptake was recently reported in pumpkin and these cultivars should be useful in elucidating the uptake mechanism.

8.3.3. Form and distribution of Si in the plant

Silicon is taken up by the roots in the form of silicic acid. After uptake, Si is translocated to the shoot together with transpiration stream and then polymerized and accumulated on the cell wall. More than 90% of Si in the plant is present in the form of silica gel. Yoshida developed an HF etching method to observe the localization of Si in rice tissues. He observed the formation of silica-cuticle double layer and silica-cellulose double layer on the cell surface in rice. The formation of these layers was suggested to prevent excessive transpiration and increase resistance to diseases and pests, which are important for the healthy growth of rice.

Silicon also accumulates on the bulliform cells, dumbbell cells, long and short cells on the surface of leaves and hulls, where are the end of transpiration stream. The silicification proceeds from dumbbell cells to bulliform cells. In roots, Si is distributed uniformly. These results indicate that distribution of Si in rice is closely related with transpiration. Therefore, the high Si content in rice shoot can be attributed to the specific and high uptake of Si, easy translocation of Si together with the transpiration stream, concentration at the end of the stream, and automatical polymerization property at a high concentration of silicic acid.

According to the chemistry of Si, silicic acid polymerizes at a concentration higher than 2 mM at 25°C. However, the Si concentration in the xylem sap of rice sometimes exceeds 2 mM. This suggests that Si may combine with some organic matter to prevent the polymerization during the translocation. However, the form of Si in the xylem sap has not been identified. Furthermore, the mechanisms of binding between Si and cell wall, Si and other metals such as Mn, Fe, and Al are unknown. Elucidation of these points may be helpful to understand the mechanism of the alleviation of the damage caused by the excess of a metal.

8.3.4. Beneficial effects of Si on crop growth

Numerous studies on the significance of Si in crop cultivation have been carried out, mostly focusing on rice in Japan and great achievements have been made. Since rice is a typical Si-accumulator, it is easy to observe the effect of Si deficiency on the growth. Furthermore, the cultivation system with heavy application of nitrogen and at a high density in Japan provides an environment that makes the effect of Si deficiency appear easily.

The beneficial effects of Si on the rice growth are mostly attributable to the characteristics of the silica gel accumulated on the epidermal tissues. This accumulation helps alleviate water stress by decreasing transpiration, to improve light interception characteristics by keeping leaf blade erect, to increase resistance to diseases and pests and lodging, and so on. These beneficial effects of Si are expressed obviously under the cultivation system with heavy application of nitrogen and high density. For this reason, Si is recognized as an "agronomically essential element" in Japan.

The mechanisms of Si-promoted plant growth include many factors. But one characteristic of the effect of Si is that it affects the growth during the reproductive growth stage more significantly than that during the vegetative growth stage not only in Si-accumulators such as rice but also in Si-excluders such as tomato, and intermediate type such as cucumber. The percentage of ripened spikelets in rice and barley and pollen fertility in soybean, cucumber, strawberry, and tomato are significantly reduced by the deficiency of Si. These facts indicate that Si may be involved in the process of reproduction although the mechanism remains unknown.

The beneficial effects of Si are usually expressed more clearly when plants are under various abiotic and biotic stresses. The alleviative effects of Si on abiotic and biotic stresses have been observed in numerous studies. The first studies on Si in Japan started from the interaction between Si and rice blast. It is clear that Si is effective in controlling rice blast, rust, stem borer, cucumber powder mildew, and so on. The protective effects of Si on diseases and pests have drawn attention recently because there is a high concern about the application of pesticides and fungicides. Both organic materials such as rice straw and inorganic materials such as slag may be useful as a Si sources, but it is necessary to investigate the effect of the application of these Si materials can actually reduce the rate of application of pesticides and fungicides in the field.

In most cases, the effect of Si on conferring the resistance to abiotic stress is observed in plant species containing a large amount of Si in the shoots. Silicon deposited on the tissue surface has been suggested to function as a physical barrier. It prevents physical penetration and/or makes the plant cell less susceptible to enzymatic degradation by fungal pathogens. Therefore, for the crops of Si-excluders, an approach for genetical modification of these crops to accumulate Si in the top, may be beneficial. One way is probably to introduce Si transporter of rice into these crops. In practice, a foliar application of silicate fertilizers has been reported to be effective in inhibiting powdery mildew development in cucumber, muskmelon, and grape (Menzies et al. 1992; Bowen et al. 1992). Silicon applied to leaves may deposit on the surface of leaves and play a role similar to Si taken up from the roots.

Silicon is also effective to alleviate abiotic stresses such as climate stress, metal toxicity, salt stress, nutrient unbalance stress and so on. The interaction between Si and P has been intensively investigated in Japan. It have been demonstrated that Si improves the availability of internal P by decreasing Mn and Fe uptake, rather than by improving the availability of soil P and P-absorbing capacity of the plants when P supply is limited. One the other hand, when P supply is high, Si decreases the P uptake either in Si-accumulator such as rice or in Si non-accumulators such as cucumber, strawberry, soybean, and tomato. The uptake system for P is considered to differ from that for Si from a metabolic inhibitor study. Therefore it is unlikely that Si competes with P in terms of uptake. Since this phenomenon occurs only in the solution with a high P concentration, there are two possibilities for Si-induced decrease in P uptake. The first is that P uptake is decreased by Si-induced reduction of transpiration. At a high P concentration, a part of P may be taken up through the apoplastic pathway. Therefore, the uptake may be affected by the transpiration. The second possibility is that Si accumulated on the roots affects the apoplastic flow of P. Because the Si-induced decrease in P uptake occurs in both Si-accumulators and non-accumulators, the second possibility is more likely. However, the exact mechanism remains unknown.

The alleviative effect of Si on Fe and Mn toxicity in rice has been attributed to the increased oxidizing capacity of the roots, and thereby to decreased uptake. In pumpkin and barley, different mechanisms seem to exist. In barley, Si causes homologous distribution of Mn in the leaves, while in pumpkin Si causes local distribution of Mn in trichomes. However, the mechanisms for the detoxification of excessive Mn by Si need to be studied further.

Silicon alleviates Al-induced inhibition of root elongation. The mechanism for this effect of Si has been suggested to form Al-Si complex, a non-toxic form of Al.

In hydroponically cultured tomato, the symptoms of Si deficiency are altered by the environmental factors such as temperature, and light and the P concentration in the nutrient solution. Although it was argued that the deficiency symptoms are caused by Zn deficiency, the exact mechanisms need to be re-examined.

As stated above, Si effects are more obvious under stress conditions. Plants are exposed to various biotic and abiotic stresses in the field during the growth period. Therefore, Si in plants certainly plays an important role in maintaining healthy growth. However, because Si is abundant in soil, its role in plant growth has been underestimated. As the global environment changes, the role of Si will become more and more important for high and sustainable production of crops in future.

Chapter 9

Silicon research in the world

Research on Si has also been done in many countries besides Japan in the fields of agriculture, physiology, biology, biochemistry, and so on. Several excellent review articles on Si in plant and in agriculture have been published recently. For example, Prof. Epstein at University of California, Davis contributed a review on anomaly of silicon in plant biology in 1994 and another review on Si in 1999. Savant et al. (1997) published a review titled "Silicon management and sustainable rice production". The first international symposium on "Silicon in Agriculture" was held in Florida, USA in September, 1999. The proceedings of this conference were published from Elsevier Science in 2001. The second international symposium on Si in agriculture will be held in Japan in 2002. These events reflect the increasing global concern on Si in plant growth. In this chapter, recent trends on Si research in the world are briefly introduced.

9.1. EFFECT OF SILICON ON CROP PRODUCTION

Although soils, climates, crops, cultivation systems and so on differ greatly with country, Si has been found to have beneficial effects on crop growth. A few examples are introduced here.

9.1.1. Rice

In Florida of USA, where organic and sand soils predominate, soils contain only a small amount of Si available to plants. In this region, application of calcium silicate slag resulted in a 30-plus increase in rice grain yield (Table 9.1, Snyder et al., 1986). The fertilized rice accumulated considerably more Si than unfertilized rice, but the tissue concentration of other nutrients (N, P, K, Ca, Mg, Fe, Mn, Zn, and Cu) and soil pH remained essentially unchanged. Thus, Si in the calcium silicate slag was concluded to have increased the yield. The application rate in Florida is usually 5 tons of slag per hectare (equivalent to ca. 1 ton Si/ha). A preliminary soil test has been developed to determine if Si applications would be beneficial (Snyder, 1991). When 0.5 M acetate-extractable Si is near 25 mg Si/L in soil then the Si content of rice straw may be 3% or greater and Si application is not needed. However, if

Table 9.1
Effect of calcium silicate slag on rice grain yield (adopted from Snyder et al., 1986)

Application rate (ton/ha)	Yield (ton/ha)				
	First crop			Ratoon crop	
	1981	1982	1983	1981	1983
0	3.4	5.3	4.3	1.7	2.9
2.5		6.3			
5.0		6.6	6.3	2.0	3.9
10.0	4.6	7.2			
20.0		7.7	6.9		4.4

extractable Si is 10 mg Si/L or less in soil, Si application is necessary. This criterion for Si application is lower than that in Japan (refer to Chapter 4).

After World War II, Korea suffered from a shortage of staple rice grain. Agronomists made various efforts to improve the rice production per unit area under limited agricultural land area and fertilizer supply. They found that application of ground wollastonite was effective in improving rice growth under a balanced supply of N, P, and K in more than 90% of Korean paddy soils containing less than 130 mg available SiO_2 per kg top soil (Park, 2001). Based on field trials, application of 2 tons of wollastonite per hectare was found to be of prime importance for rice production. When the available SiO_2 content in the soil increased to 130 mg/kg, the effect of Si application decreased. This critical value for Si effect is similar to that in Japan.

Rice is also the most important crop in China. The cultivation area of rice is about 30 million hectares and about 5 to 6 million hectares of paddy soils are estimated to be deficient in available Si. A lot of studies have been conducted by Chinese researchers on the effect of Si on rice production and they have been summarized by Wang et al. (2001). Rice yield responses to Si fertilization from some selective field trials range from 0 to 400%, depending on the severity of Si

Table 9.2
Effect of silicate fertilizer application on rice growth and yield

Application rate (kg/ha)	Number of panicle ($x10^4$/ha)	Number of spikelets per panicle	Percentage of ripening	Weight of 1000-grain	Yield (ton/ha)
0	4.84	74.7	85.9	22.9	7.01 (100)
75	4.94	73.9	90.9	23.7	7.87 (113)
105	5.03	74.8	91.4	23.8	8.16 (116)
135	5.03	76.8	91.5	25.5	8.23 (117)

Table 9.3
Effect of Si on rice yield at Santa Rosa, Colombia (adopted from Correa-Victoria et al., 2001)

	Yield (ton/ha)			
Si application rate (ton/ha)	0	1	2	3
Oryzica 1				
1992	2.1	3.2	3.6	3.7
1993 (Residual 92)	2.7	3.6	3.7	3.8
1993 (Residual 92+1 ton/ha Si)	4.0	4.1	3.9	4.2
Oryzica Llanos 5				
1992	4.1	4.5	5.1	5.0
1993 (Residual 92)	2.3	2.2	4.6	4.6
1993 (Residual 92+1 ton/ha Si)	4.7	4.7	4.7	4.6

deficiency. On average, over 10% increases in rice yield have been achieved with Si fertilization nationwide. Table 9.2 shows one of the results of field trials conducted in Jiangsu province (summarized by Takahashi et al., 1990). Application of a new Si fertilizer (probably sodium silicate) resulted in increase in rice yield by higher than 10%.

Because of these beneficial effects of Si on rice production, application of silicate fertilizers slags has become a routine practice in Florida, Korea, and parts of China, as in Japan.

9.1.2. Upland rice

Positive responses to Si fertilization have also been reported in upland rice. In Brazil, upland rice occupies approximately 70% of the total rice area and is planted in extensive areas in the savannas known as Cerrados *Sensu lato*. The predominant soils are Oxisols, which available Si is low. Field trials in this region indicated that application of wollastonite ($CaSiO_3$) resulted in a significant increase in rice yield (Prabhu et al., 2001). A similar response to Si fertilization was also reported in two cultivars of upland rice in Colombia (Table 9.3, Correa-Victoria et al., 2001). Furthermore, the residual applications of Si were very effective in increasing the yield of both cultivars.

9.1.3. Sugarcane

Sugarcane is also a Si-accumulator and an important economic crop. The beneficial effect of Si on sugarcane was previously observed in Hawaii. In Australia, 95% of sugarcane is grown on the narrow coastal plain that stretches along the east coast of Queensland. The soils in these regions have been producing sugarcane for up to 130 years, and apart from low levels of soluble soil Si resulting from natural weathering and leaching processes, there is

Table 9.4
Effect of Si application on sugarcane growth (adopted from Berthelsen et al., 2001)

Treatment (t/ha)	Plant height (mm) at		Soil pH
	1 week	4 week	
Control	243	709	5.0
$CaCO_3$			
2.5	235	684	5.7
5.0	252	670	6.3
$CaSiO_3$			
2.9	265	719	5.1
5.8	279	801	5.3
NaOH			
2.0	248	743	5.6
LSD 5%	34	89	

evidence of declining Si levels following long-term sugarcane production (Berthelsen et al., 2001). An experiment conducted in Australia showed that the growth of sugarcane responded to Si application well (Table 9.4, Berthelsen et al., 2001). Similar effect on sugarcane growth was also reported from southern Africa. In four out of the five field trials, significant responses, ranging from 9 to 24 tons cane/ha, were obtained in both the calcium silicate slag and lime treatments (Meyer and Keeping, 2001). On the average, the Si treatments were 5% better than the lime treatments. The increase in the yield by applying calcium silicate was associated with an increase in the Si content of the plant. In Brazil, on sandy soils, sugarcane has shown consistent responses to Si fertilization. The increase in yield, by applying cement and calcium silicate to the soils with a low Si content, ranged from 7 to 12% (Table 9.5, Korndorfer and Lepsch, 2001).

Table 9.5
Effect of calcium silicate on sugarcane yield cultivated on sandy soil in Brazil (adopted from Korndorfer and Lepsch, 2001)

Calcium silicate (kg/ha)	Yield (t/ha) at	
	Barreiro Farm	Amoreira Farm
0	145	128
700	153	134
1400	154	136
2800	163	137
5600	161	135

Table 9.6
Effect of Si supply (1.0 mM) on the yield of rockwool-grown cucumber in two commercial greenhouses in two successive years (adopted from Voogt and Sonneveld, 2001)

Grower		-Si		+Si	
		kg m^{-2}	Fruit wt. (g)	kg m^{-2}	Fruit wt. (g)
1	1st year	58.8	485	62.2	492
1	2nd year	67.8	510	70.9	505
2	1st crop	28.2	520	28.5	544
2	2nd crop	26.6	498	27.6	485
2	3rd crop	10.9	-	11.3	-

9.1.4. Horticultural crops

Most horticultural crops belong to non Si-accumulators. The reports on Si deficiency symptoms in tomato and cucumber by Miyake and Takahashi (1978, 1983) have generated great interest on Si nutrition in horticultural crops. In the Netherlands, where soilless culture prevails, it was found that the Si contents in plant tissue grown in soilless media were significantly lower in comparison with crops grown in soil (Voogt and Sonneveld, 2001). By adding Si to the root environment, the Si contents were increased in cucumber, melon, courgette, strawberry and bean, but not in tomato, sweet pepper, lettuce, gerbera, and carnation. The positive effect of adding Si on the yield of cucumber, courgette and rose was also obtained (Table 9.6, Voogt and Sonneveld, 2001). Nowadays, addition of Si to the nutrient solution at 0.75 mM in the form of potassium silicate is recommended.

The beneficial effects of Si on plant growth have also been observed in many other plants species, such as cotton, wheat, melon, cocoa, peanuts, apple and tobacco, in addition to the plant species described above.

9.2. ROLE OF SILICON IN DISEASE AND PEST CONTROL

It is well known that Si application reduces the severity of some diseases. The increased rice yield brought by Si application as described above is mostly attributed to Si-induced suppression of diseases and pests. This function of Si has generated great interest in the world because this implies that Si application can reduce the application of pesticide and fungicide, therefore reducing environmental risk.

Silicon has been reported to be effective for controlling various diseases and pests in different plant species. A list of diseases whose occurrence is suppressed by Si is shown in Table 9.7. Rice blast caused by the fungus *Magnaporthe grisea* (Hebert) Barr is one of the most serious threats to rice

Table 9.7
Diseases which occurrence is suppressed by Si

Crops	Disease
Rice	Blast
	Sheath bright
	Brown spot
	Stem rot
	Leaf scald
	Grain discoloration
Barley	Powdery mildew
Wheat	Powdery mildew
Cucumber	Powdery mildew
	Stem rot
	Root-infecting fungi (*Pythium* spp.)
Muskmelon	Powdery mildew

production in the world. A number of studies have shown that Si confers resistance to rice blast. In Florida, where soil is deficient in Si, application of silicate fertilizer is as effective as a fungicide in controlling rice blast (Figure 9.1, Datnoff et al. 1997).

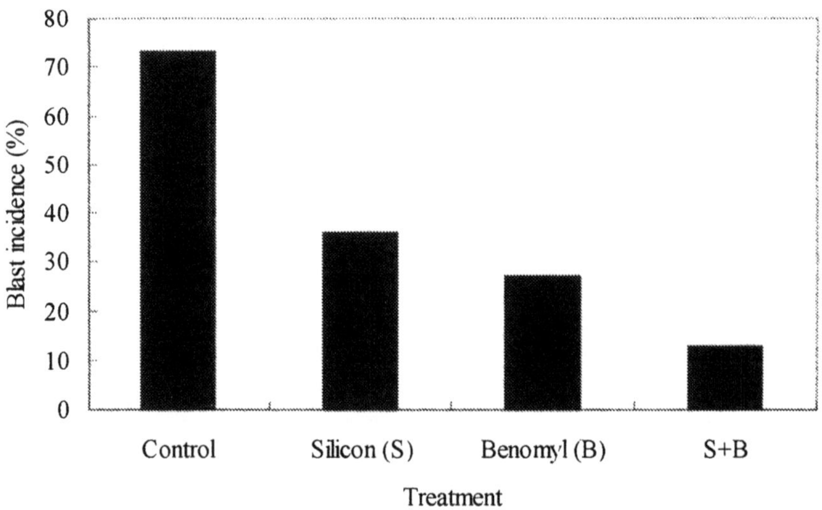

Figure 9.1. Influence of silicon fertilization (2 tons slag/ha) and foliar spray of benomyl (1.68 kg/ha) on blast incidence (adopted from Datnoff et al., 1997).

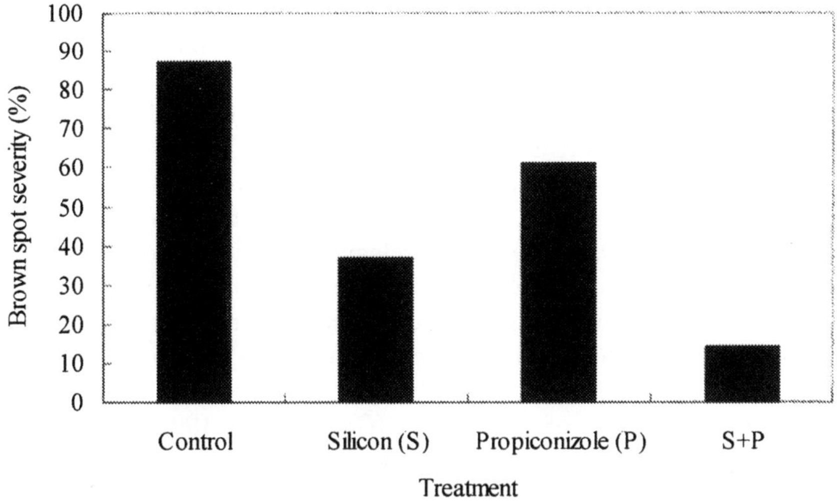

Figure 9.2. Effect of silicon (2 ton slag/ha) and propiconizole (0.44 L/ha) on brown spot severity. Brown spot severity based on a scale of 0-9, where 0=no disease and 9=76% or more of leaf area affected (adopted from Datnoff et al., 1997).

In case of brown spot, Si alone was more effective than propaconizole (Figure 9.2, Datnoff et al., 1997). For both diseases, the greatest disease control was obtained by using both treatments together. Thus Si application provides

Table 9.8
Effect of Gly extracts[a] from roots of cucumber plants grown in nutrient solutions amended with (+Si) or without (-Si) soluble silicon on oospore germination of *Pythium aphanidermatum* (adopted from Cherif et al., 1994)

Concentration	Percent inhibition[b]	
(μl/ml)	-Si	+Si
0.2	15	47
0.4	27	64
0.8	48	75
1.2	50	77
1.6	47	87
2.0	56	92

[a]Aglycones of glycosidically bound phenolics (Gly) obtained, after acid hydrolysis, from root tissues collected 6 days after inoculation with *P. aphanidermatum*.
[b]Germination of 100-200 oospores per plate after 12 h at 25°C.

control for two economically important diseases while currently available fungicides used do not have the same broad spectrum of activity.

Silicon has also been reported to prevent powdery mildew and stem rot diseases of cucumber, which is not a Si-accumulating plant (Voogt and Sonneveld, 2001). Silicon is also effective in increasing the resistance to fungal pathogen in cucumber roots (Cherif et al. 1994).

Foliar application of Si has been reported to be effective in inhibiting powdery mildew development on cucumber, muskmelon, and grape leaves (Menzies et al.,1992; Bowen et al., 1992). Si applied to leaves may deposit on the surface of leaves and play a role similar to Si taken up from the roots. This approach may be useful for the crops showing a passive or rejective Si uptake.

Silicon deposited on the tissue surface functions as a physical barrier as described in previous chapter, but recently it is postulated that Si functions as a signal to induce production of phytoalexin (Cherif et al. 1994). Si applied to cucumber stimulates chitinase activity and rapidly activates peroxidases and polyphenoloxidases after infected with *Pythium* spp. Glycosidically bound phenolics extracted from Si-treated plants after acid or beta-glucosidase hydrolysis display a strong fungistatic activity (Table 9.8).

Silicon also suppresses insect pests such as the stem borer, brown plant-hopper, rice green leafhopper, and whitebacked planthopper, and noninsect pests such as leaf spider and mites (Savant et al. 1997). In a field study, a positive relationship between Si content of rice and resistance to brown planthopper has been observed (Sujatha et al. 1987). These protective functions of Si may be attributed to Si deposited on the tissue surface.

9.3. ALLEVIATIVE EFFECT OF SILICON ON ABIOTIC STRESSES

Another recent interest on Si is its alleviative effect on various abiotic stresses. Alleviative effect of Si on Al toxicity has been observed in sorghum, barley, teosinte, maize, rice, and soybean (for reviews see Hodson and Evans, 1995; Cocker et al. 1998). Several mechanisms for the alleviative effect of Si have been proposed including solution effect (coprecipitation or inactivation of Al in the media), codeposition of Al with Si within the plant, action in the cytoplasm, effect on enzyme activity and indirect effects. The alleviative effect of Si on Al toxicity differs with plant species, probably due to differential Al tolerance and/or different mechanisms involved.

Alleviative effect of Si on Mn toxicity has also been intensively studied in the world. Although rice and pumpkin have been mainly used as experimental materials in Japan, barley (Williams and Vlamis 1957) and bean (Horst and Marschner 1978) have been used in other countries. In bean (Horst and Marschner 1978) and barley (Williams and Vlamis 1957), Si does not reduce Mn uptake, but causes homogenous distribution of Mn in the leaf blade. Although the mechanism for this homogenous distribution is still unknown, Horst et al. (1999) found that Si led to a lower apoplastic Mn concentration in cowpea and suggested that Si modifies the cation binding properties of cell wall. However, recently, further research by the same group indicated that the maintenance of the reduced state of the apoplast by soluble Si is also involved in the Si-enhanced Mn tolerance in cowpea (Iwasaki et al., 2002a, b). This is supported by the evidence that there was no correlation between apoplastic Mn concentration and the expression of Mn toxicity, but there was a negative correlation between the apoplastic Si concentration and the expression of Mn toxicity (Figure 9.3). A negative correlation was observed between the apoplastic guaiacol peroxidase (POD) activity and the Si concentrations in apoplastic washing fluid (AWF) (Figure 9.4). Silicon seems to affect the oxidation process of excess Mn mediated by POD through the interaction with phenolic substances in the solution phase of the apoplast (Iwasaki et al., 2002a).

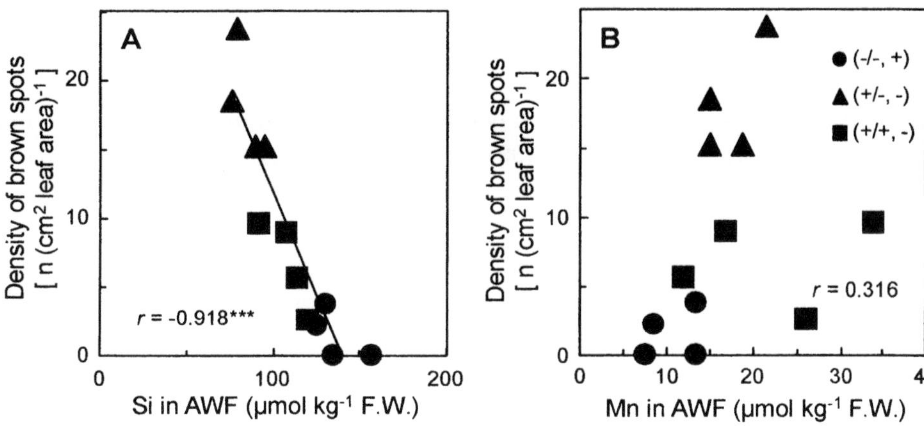

Figure 9.3. Relationships between Si (A) or Mn (B) concentration in the apoplastic washing fluid (AWF) and densities of brown spots. ●(-/-, +): plants precultured for 20 days in 0.2 μM Mn solution without Si, were treated with 50 μM Mn for 7 days in the presence of 1.43 mM Si. ■(+/+, -): plants precultured for 20 days in 0.2 μM Mn solution containing 1.43 mM Si, were treated with 50 μM Mn for 7 days in the absence of Si. ▲(+/-, -): plants were treated in the same way as those in the (+/+, -) group, except that Si supply during preculture ended 3 days prior to the beginning of the Mn treatment. The AWF was extracted with 50 mmol/L MES-Tris (pH 6.5) (adopted from Iwasaki et al., 2002a).

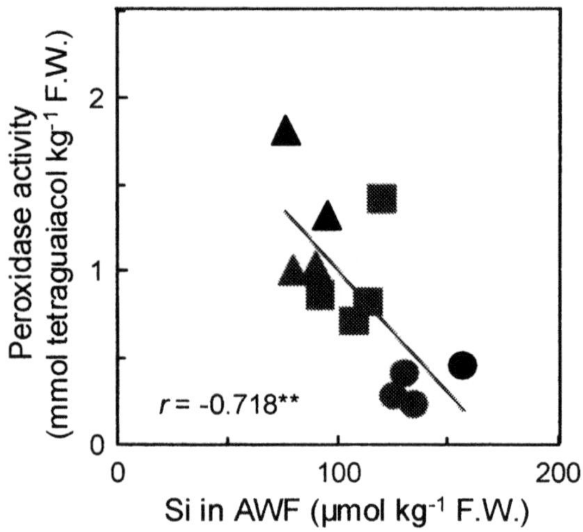

Figure 9.4. Relationship between Si concentrations in the apoplastic washing fluid (AWF) and peroxidase activities in the AWF. Refer to footnote of Figure. 9.3 for symbols (adopted from Iwasaki et al., 2002a).

In addition to Al and Mn toxicity, a beneficial effect of Si under salt stress has been observed in rice (Yeo et al. 1999), wheat (Ahmad et al. 1992) and barley (Liang et al. 1996).

Appendix 1

SiO_2 concentration of 380 river waters in 5 regions of Japan (after Kobayashi, 1961)

Region	Prefecture	Number of rivers tested	SiO_2 concentration (ppm) Maximum	Minimum	Average
I	Hokkaido	40	49.7	10.2	27.0
II	Aomori	14	49.6	12.1	27.5
	Iwate	21	28.1	13.2	18.3
	Miyagi	12	38.7	19.6	27.1
	Akita	10	26.1	15.6	18.4
	Yamagata	14	33.1	9.6	17.2
	Fukushima	11	30.7	14.3	22.0
	Ibaragi	7	26.3	13.8	21.6
	Tochigi	8	35.4	13.0	23.9
	Gunma	13	61.5	10.5	33.3
	Saitama	7	27.8	10.8	15.0
	Chiba	4	40.2	21.1	30.8
	Tokyo	5	22.8	13.7	17.0
	Kanagawa	5	33.0	20.4	27.3
	Yamanashi	5	45.2	12.4	27.5
	Nagano	16	44.1	9.1	19.7
	Niigata	14	24.8	8.0	14.2
	16 Prefectures	166	61.5	8.0	21.9
III	Toyama	6	13.3	9.9	11.4
	Ishikawa	6	27.7	9.9	14.4
	Gifu	6	13.3	7.6	11.0
	Shizuoka	8	37.7	9.9	20.7
	Aichi	4	14.5	12.3	13.6
	Mie	4	14.9	11.0	12.5
	6 Prefectures	34	27.7	7.6	14.4

Region	Prefecture	Number of rivers tested	SiO_2 concentration (ppm) Maximum	Minimum	Average
IV	Fukui	8	11.9	7.4	9.6
	Shiga	3	13.9	4.1	9.6
	Kyoto	4	15.4	5.0	10.9
	Osaka	2	20.2	7.2	13.7
	Hyogo	6	14.7	12.2	12.9
	Nara	1			20.2
	Wakayama	5	14.9	11.0	13.2
	Tottori	7	31.7	10.3	19.7
	Shimane	5	18.1	12.9	14.7
	Okayama	8	15.2	12.6	13.6
	Hiroshima	6	18.6	10.9	15.5
	Yamaguchi	6	15.6	12.5	13.8
	Tokushima	3	9.0	7.7	8.5
	Kagawa	4	20.4	17.3	18.4
	Ehime	5	17.4	6.7	12.7
	Kouchi	3	11.3	9.2	10.0
	16 Prefectures	76	31.7	4.1	13.6
V	Fukuoka	7	43.8	19.0	23.6
	Saga	5	23.1	14.1	20.1
	Nagasaki	3	47.8	14.1	24.9
	Kumamoto	13	50.4	12.2	30.3
	Ohita	7	54.6	15.5	41.7
	Miyazaki	17	50.6	10.9	28.2
	Kagoshima	12	52.0	16.8	39.1
	7 Prefectures	64	54.6	10.9	30.9
I-V	46 Prefectures	380	61.5	4.1	21.6

Appendix 2

Results of nation-wide survey on SiO_2 contents in flag leaf of rice plants conducted by Ministry of Agriculture, Forestry and Fisheries in 1955

Region	Prefecture	Number of sampling sites	SiO_2 % of flag leaf	Distribution percentage of SiO_2 contents[a]				
				A	B	C	D	E
I	Hokkaido	278	13.6	2.9	39.2	42.1	13.6	2.2
II	Aomori	1,744	16.0	3.5	18.5	39.5	29.6	8.9
	Iwate	1,072	20.0	0.3	4.2	24.9	42.4	28.2
	Miyagi	914	18.7	-	5.2	31.7	48.1	15.0
	Akita	1,338	17.7	0.8	8.5	38.7	39.9	12.0
	Yamagata	868	18.4	0.7	10.7	30.1	34.7	23.8
	Fukushima	1,186	19.8	-	3.2	22.9	49.3	24.6
	Ibaragi	530	15.7	2.4	29.1	39.1	18.7	10.7
	Tochigi	665	17.3	0.6	8.0	43.3	41.2	7.0
	Gunma	621	20.6	-	0.1	10.0	68.1	21.8
	Saitama	1,155	18.5	-	3.3	37.4	48.7	10.6
	Chiba	2,102	17.9	1.2	14.0	29.8	32.2	22.7
	Tokyo	225	14.4	-	31.5	52.4	15.6	0.4
	Kanagawa	417	20.8	-	0.5	12.0	61.9	25.6
	Yamanashi	884	16.9	0.5	11.3	47.2	34.1	6.9
	Nagano	1,381	18.1	0.9	15.4	28.1	36.3	19.3
	Niigata	800	13.2	7.7	32.7	39.6	16.5	3.5
	16 Pref.	15,902	17.8	1.2	12.3	32.9	38.6	15.1
III	Toyama	1,441	15.2	7.5	27.5	28.9	26.2	9.9
	Ishikawa	599	13.4	3.8	34.9	51.0	10.0	0.3
	Gifu	1,716	9.8	27.5	50.6	20.5	1.3	0.1
	Shizuoka	1,106	11.9	8.4	54.4	30.1	6.6	0.5
	Aichi	2,243	11.9	8.5	51.2	35.3	4.5	0.5
	Mie	954	15.7	1.1	23.2	44.1	27.4	4.2
	6 Pref.	8,059	13.0	9.5	40.3	35.0	12.7	2.5

Region	Prefecture	Number of sampling sites	SiO_2 % of flag leaf	Distribution percentage of SiO_2 contents[a]				
				A	B	C	D	E
IV	Fukui	630	10.9	11.6	60.3	28.1	-	-
	Shiga	1,455	13.4	15.5	31.0	33.7	15.4	4.4
	Kyoto	317	12.4	5.7	45.8	37.5	9.7	1.3
	Osaka	450	13.3	2.7	41.5	45.1	9.1	1.6
	Hyogo	733	12.5	17.6	29.3	38.8	14.0	0.3
	Nara	302	16.8	0.3	11.3	45.4	39.0	4.0
	Wakayama	408	11.1	29.1	34.3	22.3	11.3	3.0
	Tottori	382	14.6	2.6	17.5	59.7	19.4	0.8
	Shimane	385	15.3	0.4	22.6	45.2	28.3	3.4
	Okayama	470	9.9	28.7	50.7	17.4	3.2	-
	Hiroshima	323	9.5	20.7	68.7	10.6	-	-
	Yamaguchi	692	11.9	3.0	51.8	40.9	4.3	-
	Tokushima	583	15.1	2.1	23.3	48.1	25.2	1.4
	Kagawa	694	16.2	-	10.9	55.6	32.6	0.9
	Ehime	766	15.7	4.2	24.0	37.5	24.4	10.9
	Kouchi	512	11.9	10.1	57.2	29.1	3.2	0.4
	16 Pref.	9,102	13.2	9.6	36.3	37.2	14.9	2.0
V	Fukuoka	1,014	16.6	-	13.0	44.1	35.0	7.9
	Saga	246	18.3	0.8	11.4	48.8	29.3	9.7
	Nagasaki	203	13.9	3.9	28.1	51.3	14.3	2.4
	Kumamoto	946	16.1	1.9	17.9	41.9	32.2	6.1
	Oita	680	21.5	-	2.2	13.7	41.1	43.0
	Miyazaki	902	17.1	0.7	14.7	37.4	37.4	9.8
	Kagoshima	617	18.8	0.8	3.7	26.4	57.6	11.5
	7 Pref.	4,608	17.5	1.2	13.0	37.6	35.3	12.9
I-V	46 Pref.	37,949	15.6	5.4	24.2	34.3	26.7	9.4

A, <7.5%; B, 7.6-12.5%; C, 12.6-17.5%; D, 17.6-22.5%; E, >22.6%

Appendix 3-A

Content of Si and Ca in Angiospermae, Gymnospermae, Pteridophyta and Bryophyta collected from Nippon Shinyaku Institute for Botanical Research

	Si %	Ca %	Si/Ca	Degree of Si accumulation
MAGNOLIOPHYTA(ANGIOSPERMAE)				
MAGNOLIOPSIDA(DICOTYLEDONEAE)				
Magnoliales				
Magnoliaceae				
Magnolia gradiflora L.	0.67	1.80	0.37	±
Cercidiphyllaceae				
Cercidiphyllum japonicum Sieb.et Zucc.	0.94	2.65	0.35	±
Lauraceae				
Lindera strychnifolia(Sieb.et Zucc.)F.Vill.	0.09	0.76	0.12	-
Piperales				
Saururaceae				
Houttuynia cordata Thunb.	1.15	1.17	0.98	±
Saururus chinensis(Lour.)Baill.	0.28	1.79	0.16	-
Ranunculales				
Ranunculaceae				
Aconitum japonicum Thunb.	0.35	3.97	0.09	-
Aconitum loczyanum Rapaics	0.32	4.14	0.08	-
Aquilegia flabellata Sieb.et Zucc.	0.05	1.96	0.03	-
Ranunculus japonicus Thunb.	0.36	3.36	0.11	-
Berberidaceae				
Epimedium grandiflorum Morr.	0.39	2.36	0.17	-
Nandina domestica Thunb.	0.19	0.74	0.26	-
Theales				
Theaceae				
Camellia japonica L.	0.06	1.77	0.03	-
Camellia sasanqua Thunb.	0.15	0.86	0.17	-
Thea sinensis L.	0.04	1.00	0.04	-
Malvales				
Malvaceae				
Hibiscus syriacus L.	0.18	3.38	0.05	-

	Si %	Ca %	Si/Ca	Degree of Si accumulation
Papaverales				
Papaveraceae				
Papavera bracteatum Lindl.	0.75	1.32	0.57	±
Papavera rhoeas L.	0.58	3.16	0.18	±
Brassicaceae				
Armoracia rusticana P.Gaertn.	0.04	3.24	0.01	-
Wasabia japonica(Miq.)Koidz.	0.14	3.53	0.04	-
Violales				
Tamaricaceae				
Tamarix chinensis Lour.	0.54	3.16	0.17	±
Violaceae				
Hybanthus glutinosus Taub.	0.27	1.17	0.23	-
Viola tricolor L.var.hortensis DC.	0.13	1.46	0.09	-
Cucurbitaceae				
Benincasa hispida (Thunb.etJ.Murr.)Cogn.	0.42	2.92	0.14	-
Ecballium elaterium(L.)A.Rich.	0.57	2.71	0.21	±
Luffa acutangula(L.)Roxb.	0.52	4.90	0.11	±
Caryophyllales				
Caryophyllaceae				
Dianthus superbus L.	0.06	1.30	0.05	-
Saponaria officinalis L.	0.27	2.51	0.11	-
Amaranthaceae				
Amaranthus viridis L.	0.15	2.06	0.07	-
Polygonales				
Polygonaceae				
Polygonum hydropiper L.	0.15	1.94	0.08	-
Hamamelidales				
Saxifragaceae				
Hydrangea macrophylla(Thunb.ex J.Murr.)Ser.	0.34	2.08	0.16	-
Philadelphus satsumi Sieb.	0.05	0.64	0.08	-

	Si %	Ca %	Si/Ca	Degree of Si accumulation
Rosales				
Rosaceae				
Chaenomeles sinensis(Thunb.)Koehne	0.29	1.16	0.25	-
Crataegus cuneata Sieb.et Zucc.	0.11	1.86	0.06	-
Kerria japonica(L.)DC.	0.33	2.41	0.14	-
Pyracantha crenulata (D.Don)M.J.Roem	0.09	1.47	0.06	-
Spiraea thunbergii Sieb.	0.16	0.67	0.24	-
Fabaceae				
Albizzia julibrissin Durazz.	0.10	1.40	0.07	-
Sophora flavescens Solander ex Aiton	0.05	2.54	0.02	-
Sophora japonica L.var.pendula Loud.	0.07	1.55	0.05	-
Wisteria brachybotrys Sieb.et Zucc.	0.24	0.80	0.30	-
Geraniales				
Geraniaceae				
Pelargonium graveolens L'Her.ex Ait.	0.30	1.50	0.20	-
Tropaeolaceae				
Tropaeolum majus L.	0.02	1.95	0.01	-
Sapindales				
Rutaceae				
Phellodendron amurense Rupr.	0.40	3.50	0.11	-
Poncirus trifoliata(L.)Raf.	0.49	1.71	0.29	-
Zanthoxylum piperitum(L.)DC.	0.36	1.30	0.28	-
Meliaceae				
Melia azedarach L. var. toosendam Makino	0.27	3.07	0.09	-
Aceraceae				
Acer ginnala Maxim.	0.53	0.86	0.62	±
Acer saccharum Marsh.	0.86	1.25	0.69	±
Celastrales				
Aquifoliaceae				
Ilex aquifolium L.	0.10	0.44	0.23	-
Scrophulariales				
Oleaceae				

	Si %	Ca %	Si/Ca	Degree of Si accumulation
Olea europaea L.	0.13	1.39	0.09	-
Cornales				
Cornaceae				
Aucuba japonica Thunb.	0.58	2.13	0.27	±
Apiales				
Apiaceae				
Conium maculatum L.	0.03	1.33	0.02	-
Araliaceae				
Aralia cordata Thunb.	0.08	2.27	0.04	-
Panax ginseng C.A.Mey.	0.20	1.72	0.12	-
Myrtales				
Thymelaeaceae				
Daphne odora Thunb.	0.09	1.15	0.08	-
Myrtaceae				
Callistemon rigidus R.Br.	0.09	0.98	0.09	-
Melastomataceae				
Melastoma candidum D.Don	0.18	2.31	0.08	-
Onagraceae				
Oenothera Lamarckiana Ser.	0.08	3.52	0.02	-
Elaeagnaceae				
Elaeagnus multiflora Thunb. var.gigantea Araki	0.16	0.84	0.19	-
Urticales				
Moraceae				
Morus alba L.	0.77	2.11	0.36	±
Morus alba L.var.pendula	0.53	1.19	0.45	±
Fagales				
Fagaceae				
Quercus suber L.	0.34	1.35	0.25	-
Salicales				
Salicaceae				
Salix matsudana var. tortuosa Vilm.	0.14	4.29	0.03	-

	Si %	Ca %	Si/Ca	Degree of Si accumulation
Euphorbiales				
Euphorbiaceae				
Mallotus japonicus(Thunb.)Muell.Arg.	0.48	1.85	0.26	-
Securinega suffruticosa(Pall.)Rehd.	0.19	2.17	0.09	-
Ericales				
Ericaceae				
Erica canaliculata Andr.	0.16	0.51	0.31	-
Rhododendron japonicum Suringer	0.42	0.98	0.43	-
Gentianales				
Apocynaceae				
Nerium oleander L.	0.18	2.76	0.07	-
Rubiales				
Rubiaceae				
Rubia tinctorum L.	0.50	2.01	0.25	-
Lamiales				
Solanaceae				
Atropa belladonna L.	0.03	1.25	0.02	-
Physalis alkekengi L.	0.10	1.42	0.07	-
Convolvulaceae				
Calystegia japonica Choisy	0.02	0.71	0.01	-
Polemoniaceae				
Phlox subulata L.	1.11	2.58	0.43	±
Bignoniaceae				
Campsis grandiflora(Thunb.)K.Schum.	0.14	0.81	0.17	-
Catalpa ovata G.Don	0.20	1.12	0.18	-
Verbenaceae				
Verbena officinalis L.	0.47	0.86	0.55	-
Lamiaceae				
Lavandura angustifolia Mill	0.54	1.62	0.33	±
Plectranthus japonicus (Burm.)Koidz.	0.07	1.64	0.04	-
Salvia officinalis L.	0.66	1.17	0.56	±
Asterales				
Asteraceae				
Artemisia absinthium L.	0.32	1.26	0.25	-

	Si %	Ca %	Si/Ca	Degree of Si accumulation
Artemisia maritima L.	0.10	1.18	0.08	-
Baccharis halimiflolia L.	0.13	1.24	0.10	-
Chamomilla recutita(L.)Rauschert	0.16	1.90	0.08	-
Chrysanthemum coronarium L.	0.18	1.29	0.14	-
Eupatorium fortunei Turcz.	0.37	1.48	0.25	-
LILIOPSIDA(MONOCOTYLEDONEAE)				
Alismatales				
Alismataceae				
Sagittaria trifolia L.f.plena Makino	0.39	1.91	0.20	-
Liliales				
Liliaceae				
Allium fistulosum L.	0.31	3.08	0.10	-
Allium fistulosum L. var. viviparum Makino	0.03	1.09	0.03	-
Aloe arborescens Mill.	0.16	7.14	0.02	-
Anemarrhena asphodeloides Bunge	0.08	1.44	0.06	-
Asparagus cochinchinensis var.pygmaeus Ohwi	0.25	0.72	0.35	-
Asparagus officinalis L.	0.27	0.83	0.33	-
Aspidistra elatior Blume	0.06	1.48	0.04	-
Convallaria majalis L.	0.53	3.73	0.14	±
Crinum asiaticum L. var. japonicum Baker	0.02	1.40	0.01	-
Heloniopsis orientalis (Thunb.)C.Tanaka	0.11	0.50	0.22	-
Hemerocallis fulva L.var.kwanso Regel	0.18	0.62	0.29	-
Hosta longissima Honda	0.09	2.32	0.04	-
Lilium leichtlinii Hook	0.11	2.08	0.05	-
Lycoris radiata Herb.	0.01	2.70	0.01	-
Polygonatum odoratum var.pluriforum(Miq.)Ohwi	0.09	1.32	0.07	-
Rohdea japonica (Thunb.)Roth.	0.25	1.15	0.22	-
Sansevieria trifasciata Hort.ex Prai	0.01	1.15	0.01	-
Smilacina japonica A.Gray	0.54	3.72	0.18	±
Tricyrtis hirta(Thunb.)Hook	0.24	2.93	0.08	-

	Si %	Ca %	Si/Ca	Degree of Si accumulation
Yucca filamentosa L.	0.08	0.48	0.17	-
Zephyranthes candida (Lindl.)Herb.	0.10	0.90	0.11	-
Agavaceae				
Agave americana L.	0.14	1.86	0.08	-
Stemonaceae				
Stemona japonica Miq.	0.23	3.45	0.07	-
Iridaceae				
Iris ensata Thunb.var.ensata Nakaki	0.15	0.95	0.16	-
Iris florentina L.	0.08	1.51	0.05	-
Iris setosa Pall.	0.18	0.46	0.39	-
Zingiberales				
Musaceae				
Musa basjoo Sieb.et Zucc.	1.08	0.75	1.44	+
Zingiberaceae				
Zingiber mioga(Thunb.)Rosc.	0.22	1.10	0.20	-
Cannaceae				
Canna indica L.	0.36	1.24	0.29	-
Orchidales				
Orchidaceae				
Bletilla striata (Thunb.)Rchb.	0.47	1.42	0.33	-
Spiranthes sinensis(Pers.)Ames	0.06	1.80	0.03	-
Arales				
Araceae				
Acorus calamus L.var.pusillus Engler	0.04	1.26	0.03	-
Acorus gramineus Soland	0.10	1.31	0.08	-
Amorphophallus rivieri Dur.	0.04	1.44	0.03	-
Pistia stratiotes L.	0.17	3.26	0.05	-
Arecales				
Arecaceae				
Phoenix dactylifera L.	0.27	0.97	0.28	-
Phoenix roebelenii O'Bien	0.30	1.23	0.24	-
Rhapis humilis Blume	0.52	1.09	0.48	±
Trachycarpus fortunei(Hook)H.Wendl.	1.41	0.88	1.60	+

	Si %	Ca %	Si/Ca	Degree of Si accumulation
Bromeliales				
Bromeliaceae				
Ananas comosus(L.)Merr.	0.24	1.45	0.17	-
Commelinales				
Juncaceae				
Juncus effusus L.var.decipiens Buch.	0.60	0.49	1.22	±
Commelinaceae				
Tradescantia ohiensis Rafin.	0.39	2.72	0.14	-
Cyperales				
Cyperaceae				
Carex biwensis Franch.	1.22	0.15	8.13	+
Carex conica Boott.	2.54	1.20	2.12	+
Carex dispalata Boott. var.dispalata.	2.41	1.00	2.41	+
Carex parciflora Boott. var.macroglossa Ohwi.	1.73	0.40	4.33	+
Carex thunbergii Steud.	1.67	0.23	7.26	+
Cyperus alternifolius L.	3.52	0.56	6.29	+
Cyperus microiria Steud.	0.91	0.52	1.75	+
Cyperus papyrus L.	1.75	0.84	2.08	+
Scirpus tabernaemontani Gmel.	0.20	0.49	0.41	-
Scirpus tabernaemontani Gmel. f. zebrinus A.et G.	0.22	0.54	0.41	-
Poaceae (Gamineae)				
Aegilops squarrosa L.	2.11	0.58	3.64	+
Arundo donax L.	1.24	0.56	2.21	+
Avena sativa L.	2.08	1.08	1.93	+
Cortaderia selloana Asch.et Graebn.	0.65	0.34	1.91	+
Cymbopogon citratus(DC.)Stapf.	0.85	0.71	1.20	+
Elymus mollis Trin	0.95	0.67	1.42	+
Miscanthus sinensis Anderss.	2.96	0.65	4.55	+
Oryza sativa L.	6.30	1.31	4.81	+
Pleioblastus chino Makino	5.17	0.53	9.75	+
Saccharum officinarum L.	0.77	0.43	1.79	+
Sasa nipponica Makino et Shibata	4.32	0.40	10.8	+

	Si %	Ca %	Si/Ca	Degree of Si accumulation
Secale cereale L.	1.04	0.80	1.35	+
Triticum aestivum L.	1.44	0.42	3.43	+
Triticum boeoticum Boiss	2.61	0.46	5.67	+
Triticum dicoccoides Schwinf.	1.33	0.66	2.02	+
Triticum percicum Vav. x Aegilops squarrosa L.	1.70	0.43	3.95	+
GYMNOSPERMOPHYTA				
Cycadopsida				
Cycadales				
Cycadaceae				
Cycas revoluta Thunb.	0.07	0.87	0.08	-
Ginkgoales				
Ginkgoaceae				
Ginkgo biloba L.	0.05	3.53	0.01	-
Coniferopsida				
Taxodiales				
Cupressaceae				
Chamaecyparis obtusa Sieb.et Zucc.	0.12	1.01	0.12	-
Cupressus sempervirens L.	0.10	1.19	0.08	-
Thuja orientalis L.	0.20	1.30	0.15	-
Taxaceae				
Torreya nucifera(L.) Sieb.et Zucc.	0.11	1.79	0.06	-
Pinaceae				
Pinus luchuensis Mayr.	0.08	0.46	0.17	-
Pinus palustris Mill.	0.33	0.41	0.80	-
Taxodiaceae				
Cryptomeria japonica D.Don	0.14	1.11	0.13	-
Cunninghamia lanceolata(Lamb.)Hook	0.06	1.41	0.04	-
Sequoia sempervirens(D.Don)Endl.	0.24	0.80	0.30	-
Gnetales				
Ephedraceae				
Ephedra sinica Stapf.	0.02	0.56	0.04	-

	Si %	Ca %	Si/Ca	Degree of Si accumulation
PTERIDOPHYTA				
Lycopsida				
Lycopodiales				
Lycopodiaceae				
Lycopodium clavatum L.	0.64	0.11	5.82	+
Selaginellales				
Selaginellaceae				
Selaginella caulescens Spring.	0.91	0.54	1.69	+
Selaginella involvens Spring.	2.59	0.28	9.25	+
Selaginella uncinata Spring.	1.84	1.06	1.74	+
Equisetopsida				
Equisetales				
Equisetaceae				
Equisetum arvense L.	4.25	4.94	0.86	+
Equisetum hiemale L.	2.48	1.88	1.32	+
Filicopsida				
Filicales				
Pteridaceae				
Adiantum pedatum L.	0.53	1.10	0.48	±
Aspleniaceae				
Asplenium trichomanes L.	0.09	0.56	0.16	-
Blechnaceae				
Blechnum amabile Makino	0.93	1.32	0.70	±
Struthiopteris niponica Nakai	3.17	0.84	3.77	+
Polypodiaceae				
Loxogramme saziran Tagawa ex Price	0.12	0.52	0.23	-
Pyrrosia lingua(Thunb.)Farwell	0.05	1.00	0.05	-
Schizaeaceae				
Lygodium japonicum(Thunb.)Sw.	1.20	0.23	5.22	+
Gleicheniaceae				
Gleichenia glauca Hook	0.79	0.12	6.58	+

	Si %	Ca %	Si/Ca	Degree of Si accumulation
BRYOPHYTA				
Bryopsida				
Sphagnales				
Sphagnaceae				
Sphagnum cymbifolium Warnst.	1.37	1.07	1.28	+
Hepaticopsida				
Marchantiales				
Marchantiaceae				
Marchantia polymorpha L.	5.55	1.04	5.33	+

Appendix 3-B

Content of Si and Ca in Pteridophyta collected from Kyoto Prefectural Botanical Garden

	Si %	Ca %	Si/Ca	Degree of Si accumulation
PTERIDOPHYTA				
Lycopsida				
Selaginellales				
Selaginellaceae				
Selaginella caulescens Spring.	5.29	0.65	8.14	+
Selaginella involvens Spring.	3.91	0.45	8.69	+
Equisetopsida				
Equisetales				
Equisetaceae				
Equisetum arvense L.	6.00	2.67	2.25	+
Equisetum hiemale L.	5.62	1.69	3.33	+
Filicopsida				
Marattiales				
Marttiaceae				
Angiopteris lygodiifolia Rosenst.	1.66	1.23	1.35	+
Filicales				
Osmundaceae				
Osmunda japonica Thunb.	5.59	1.56	3.58	+
Osmunda lancea Thunb.	2.42	0.51	4.75	+
Blechnaceae				
Woodwardia orientalis Swartz.	2.35	1.11	2.12	+
Pteridaceae				
Adiantum pedatum L.	2.20	0.78	2.82	+
Dennstaedtia scabra (Moore) Chr.	1.90	0.59	3.22	+
Pteridium aquilinum Kuhn	4.96	1.26	3.94	+
Pteris ensiformis Burm.	1.63	0.74	2.20	+
Thelypteridaceae				
Lastrea oligophlebia E.Copel.	1.22	2.25	0.54	±
Cyclosorus acuminatus (Houtt.) Nakai	3.27	1.77	1.85	+

	Si %	Ca %	Si/Ca	Degree of Si accumulation
Cyclosorus dentatus (Forssk.) Ching	5.11	1.69	3.02	+
Leptogramma mollissima(Fisch.) Ching	3.00	2.19	1.37	+
Athyriaceae				
Athyrium japonicum (Thunb.) Copel	1.22	1.38	0.88	+
Athyrium lobato-crenatum Tagawa	1.91	1.34	1.43	+
Athyrium niponicum (Mett.) Hance	0.97	0.82	1.18	+
Athyrium yokoscense (Fr.et Sav.) Christ	1.43	1.56	0.92	+
Diplazium wichurae (Mett.) Diels	1.67	1.01	1.65	+
Diplazium hachijoense Naki	4.39	2.55	1.72	+
Onoclea sensibilis L. var. interrupta Maxim.	1.54	2.00	0.77	+
Davalliaceae				
Davallia mariesii Moore ex Bak.	0.46	0.64	0.72	-
Nephrolepis cordifolia(L.)K.Presl	0.27	1.84	0.15	-
Polypodiaceae				
Colysis decurrens Nakaike	0.30	1.03	0.29	-
Colysis wrightii (Hook.) Ching	0.03	0.68	0.04	-
Pyrrosia lingua (Thunb.) Farw.	0.04	0.69	0.06	-
Dryopteridaceae				
Acrophorus stipellatus (Wall.) Moore	1.13	2.11	0.54	±
Ctenitis subglandulosa (Hance) Ching	0.49	1.65	0.36	-
Cyrtomium falcatum (L.f.) K.Presl	0.13	0.96	0.14	-
Cyrtomium fortunei J.Sm.	0.23	1.46	0.16	-
Dryopteris bissetiana (Bak.) C. Chr.	0.06	1.12	0.05	-

	Si %	Ca %	Si/Ca	Degree of Si accumulation
Dryopteris crassirhizoma Nakai	0.14	2.06	0.07	-
Dryopteris erythrosora (Eat.) O.Kuntze	0.21	0.96	0.22	-
Dryopteris lacera (Thunb.) O.Kuntze	0.19	1.11	0.17	-
Dryopteris sieboldii O.Kuntze	0.27	2.14	0.13	-
Dryopteris uniformis Makino	0.24	1.37	0.18	-
Polystichopsis amabilis (Blume) Tagawa	0.33	1.33	0.25	-
Polystichopsis pseudo-aristata Tagawa	0.29	1.34	0.22	-
Polystichopsis standishii (Moore) Tagawa	0.21	1.58	0.13	-
Polystichum lepidocaulon J.Sm.	0.06	0.84	0.07	-
Polystichum polyblepharum (Roem.) K.Presl	0.07	0.75	0.09	-
Polystichum pseudo-makinoi Tagawa	0.37	1.81	0.20	-
Polystichum tripteron (Kunze) K.Presl	0.19	1.51	0.13	-

Appendix 3-C

Content of Si and Ca in Oryzeae collected from National Institute of Genetics

	Si %	Ca %	Si/Ca	Degree of Si accumulation
Gramineae				
Bambusoideae				
Oryzeae				
Leersia japonica Makino	6.83	1.11	6.15	+
Leersia oryzoides Sw. var. japonica Hack.	1.20	0.45	2.67	+
Oryza alta W0017P1	5.81	0.38	15.3	+
Oryza australiensis W1562P1	9.92	0.51	19.5	+
W1562P2	10.41	0.47	22.1	+
Oryza brachyantha W1407S1	6.41	0.36	17.8	+
W1407S3	7.76	0.38	20.4	+
Oryza breviligulata W1423	9.41	0.43	21.9	+
W0925	10.62	0.45	23.6	+
Oryza eichingeri W1522P3	7.02	0.39	18.0	+
W1526P1	10.01	0.58	17.3	+
Oryza glaberima W0025	8.43	0.42	20.1	+
Oryza grandiglumis W1480-8P1	5.10	0.28	18.2	+
W1483P1	4.96	0.25	19.8	+
Oryza latifolia W1178P1	7.20	0.41	17.6	+
Oryza longiglumis W1215P1	6.63	0.35	18.9	+
W1216P1	4.18	0.29	14.4	+
Oryza meyeriana W0004P3	3.42	0.82	4.2	+
W1353P1	2.78	0.72	3.9	+
Oryza minuta W1322P1	8.32	0.46	18.1	+
W1329P1	5.63	0.44	12.8	+
Oryza officinalis W1250P1	11.82	0.43	24.5	+
W1291P1	10.03	0.55	18.2	+
Oryza perennis AF011	9.70	0.33	29.4	+
AF110	6.43	0.31	20.7	+
W0036	5.64	0.38	14.8	+

	Si %	Ca %	Si/Ca	Degree of Si accumulation
Oryza perennis W0105	9.59	0.42	22.8	+
W0106	7.55	0.36	21.0	+
W0107	9.07	0.46	19.7	+
W0120	5.38	0.42	12.8	+
W0120P22	6.45	0.47	13.7	+
W0122	8.60	0.42	20.5	+
W0124	5.41	0.42	12.9	+
W0125	8.77	0.49	17.9	+
W0126	8.32	0.37	22.5	+
W0130	9.09	0.46	19.8	+
W0132	6.65	0.38	17.5	+
W0133	9.19	0.46	20.0	+
W0136	8.14	0.36	22.6	+
W0139	8.09	0.37	21.9	+
W0144	8.46	0.44	19.2	+
W0149	5.63	0.36	15.6	+
W0153	7.74	0.35	22.1	+
W0157	5.63	0.38	14.8	+
W0168	7.84	0.35	22.4	+
W0172	6.69	0.34	19.7	+
W0593	9.49	0.42	22.6	+
W0593P1	9.61	0.51	18.8	+
W0612	7.20	0.50	14.4	+
W0630	7.98	0.41	19.5	+
W1185	8.05	0.34	23.7	+
W1186	6.29	0.31	20.3	+
W1187	7.39	0.29	25.5	+
W1189	6.59	0.26	25.3	+
W1191	6.52	0.31	21.0	+
W1192	6.39	0.29	22.0	+
W1196	6.91	0.39	17.7	+
W1235	8.14	0.44	18.5	+
W1294	7.34	0.42	17.5	+
W1347P1	10.60	0.41	25.9	+

	Si %	Ca %	Si/Ca	Degree of Si accumulation
W1509P1	9.08	0.57	15.9	+
Oryza perrieri W1529P1	4.41	0.55	8.0	+
W1529P2	5.30	0.62	8.5	+
Oryza punetata W1023P2	6.91	0.45	15.4	+
W1514P1	6.02	0.58	10.4	+
Oryza ridleyi W0001S2	7.99	0.35	22.8	+
W0604P1	6.45	0.38	17.0	+
Oryza sativa T0065	8.63	0.40	21.6	+
T0108	7.74	0.36	21.5	+
Oryza subulata W0510P1	7.07	0.46	15.4	+
W0510S1	6.14	0.48	12.8	+
Oryza tisseranti W1345P1	10.88	0.71	15.3	+
Zizania latifolia Turcz.	4.40	0.65	6.77	+

Appendix 3-D

Content of Si and Ca in Bambusoideae, Pooideae, Panicoideae, Ergrostoideae collected from Kyoto Prefectural Botanical Garden.

	Si%	Ca%	Si/Ca	Degree of Si accumulation
MAGNOLIOPHYTA (ANGIOSPERMAE)				
LILIOPSIDA (MONOCOTYLEDONEAE)				
Gramineae				
Bambusoideae				
Arundinaria mikurensis Naki	2.00	0.47	4.26	+
A. sasakiana Nakai	3.88	0.46	8.43	+
Bambusa. Multiplex Raeusch.	2.90	0.56	5.18	+
B. multiplex Raeusch. var. elegans Muroi	4.16	0.74	5.62	+
B. multiplex Raeusch. f. alphonso-karii Nakai	3.71	0.66	5.62	+
B. multiplex Raeusch. f. variegata Hatsus	3.57	0.57	6.26	+
Chimonobambusa marmorea Makino	2.20	0.78	2.82	+
C. marmorea Makino f. variegata Ohwi	3.69	0.81	4.56	+
Nipponobambusa nikkoensis Muroi	6.17	1.14	5.41	+
Phyllostachys aurea (Sieb.) Carriere ex Riv.	5.58	0.57	9.79	+
P. aurea Sieb. f. albo-variegata Makino	3.28	0.59	5.56	+
P. aurea Sieb. f. alternatolutescens Makino	3.60	0.52	6.92	+
P. bambusoides Sieb. et Zucc. [Formosa]	4.07	0.50	8.14	+
P. bambusoides Sieb. et Zucc. [Marliacea]	3.64	0.63	5.78	+
P. heterocycla Mitf.	1.30	0.51	2.55	+
P. heterocycla Mitf. [Butsumenchiku]	2.10	0.51	4.12	+
P. heterocycla Mitf. [Heterocycla]	3.09	0.65	4.75	+
P. heterocycla Mitf. [Nabeshimana]	2.27	0.52	4.37	+
P. humilis Muroi	2.84	0.47	6.04	+
P. makinoi Hayata	5.27	0.73	7.22	+
P. nigra Munro	3.75	0.55	6.82	+
P. nigra Munro var. henonis Stafp	3.15	0.62	5.08	+
P. nigra Munro var. henonis f. albo-variegata Makino	3.23	0.60	5.38	+
P. nigra Munro var. henonis f. boryana Makino	3.54	0.85	4.17	+

	Si%	Ca%	Si/Ca	Degree of Si accumulation
P. nigra Munro var. henonis f. hyugaensis Makino	7.30	1.11	6.84	+
P. nigra Munro var. henonis f. megurochiku Makino	2.36	0.55	4.29	+
P. nigra Munro f. nigra-punctata Makino	3.61	0.49	7.37	+
P. nigra Munro var. tosaensis Makino	5.81	0.85	6.84	+
P. reticulata K. Koch	4.61	0.61	7.56	+
P. reticulata f. kashirodake Makino	6.13	0.90	6.81	+
P. tranquillans Muroi	3.00	0.74	4.05	+
Pleioblastus argenteostriatus Nakai	4.06	0.80	5.08	+
Pl. argenteostriatus Nakai cv. Distichus	6.70	1.11	6.04	+
Pl. chino Makino	3.82	1.26	3.03	+
Pl. chino Makino f. angustifolius Muroi	3.89	0.73	5.33	+
Pl. chino Makino var. gracilis Nakai	3.50	0.93	3.76	+
Pl. gramineus Nakai	4.29	0.71	6.04	+
Pl. hindsii Nakai	4.30	0.53	8.36	+
Pl. linearis Nakai	4.14	0.56	7.39	+
Pl. multifolius Nakai	3.91	0.92	4.25	+
Pl. simonii Nakai	4.86	1.10	4.42	+
Pl. viridistriatus Makino	5.28	0.89	5.93	+
Pl. yoshidake Nakai	3.69	0.72	5.13	+
Pseudsasa japonica Makino	4.21	0.58	7.26	+
Ps. japonica Makino cv. Tsutsumiana	3.77	0.69	5.46	+
Sasa amagiensis Makino	3.98	0.55	7.24	+
Sasa arakii Makino	2.62	0.56	4.68	+
Sasa chartacea Makino	4.47	0.81	5.52	+
Sasa glabra Koidz.	3.91	0.60	6.52	+
Sasa gracillima Nakai	7.44	0.77	9.66	+
Sasa hashimotoi Koidz.	3.23	0.71	4.55	+
Sasa hastatophylla Muroi	3.13	0.69	4.54	+
Sasa hutatabiensis Koidz.	3.61	0.53	6.81	+
Sasa kurilensis Makino et Shibata	2.64	0.41	6.44	+
Sasa kurokawana Makino	2.36	0.62	3.81	+
Sasa nikkouensis Makino et Shibata	6.91	0.65	10.63	+

	Si%	Ca%	Si/Ca	Degree of Si accumulation
Sasa nipponica Makino et Shibata	5.28	0.79	6.68	+
Sasa okadana Makino	4.63	0.80	5.79	+
Sasa paniculata Makino et Shibata	2.75	0.39	7.05	+
Sasa uyemurana Makino et Uchida	4.81	0.61	7.89	+
Sasa veitchii Rehd.	3.38	0.56	6.04	+
Sasamorpha borealis Nakai	3.85	0.47	8.19	+
Sasamorpha mollis Nakai	3.60	0.64	5.63	+
Sasamorpha uinuizoana Koidz.	4.05	0.50	8.10	+
Semiarundinaria fastuosa Makino	2.47	0.72	3.43	+
S. fastuosa Makino f. viridis Murata	3.32	0.84	3.95	+
S. kagamiana Makino	4.72	0.76	6.21	+
S. tatebeana Muroi	5.45	0.88	6.19	+
S. villosa Muroi	4.56	0.86	5.30	+
S. yoshi-matumurae Muroi	3.22	0.69	4.67	+
Shibataea kumasaca Makino	2.93	0.48	6.10	+
Sinobambusa tootsik Makino	2.56	0.76	3.37	+
S. tootsik Makino f. albostriata Muroi	3.73	1.04	3.59	+
Tetragonocalamus quadrangularis Nakai	4.65	1.10	4.23	+
Pooideae				
Agropyron kamoji Ohwi	1.83	0.33	5.55	+
Agropyron racemiferum Koidz.	2.15	0.59	3.64	+
Agrostis clavata Trin. var. nukabo Ohwi	6.97	0.67	10.40	+
Agrostis palustris Huds.	1.95	0.48	4.06	+
Arundinella hirta Tanaka	4.18	0.70	5.97	+
Arundo donax L. [Versicolor]	1.75	0.89	1.97	+
Avena fatua L.	1.71	0.91	1.88	+
Bromus pauciflorus Hack.	2.21	0.66	3.35	+
Bromus unioloides H. B. K.	2.97	1.42	2.09	+
Calamagrostis pseudo-phragmites Koeler	2.55	0.30	8.50	+
Festuca myuros L.	1.17	0.50	2.34	+
Festuca ovina L.	1.36	0.27	5.04	+
Festuca parvigluma Steud.	2.58	0.61	4.23	+
Festuca rubra L.	1.40	0.40	3.50	+

	Si%	Ca%	Si/Ca	Degree of Si accumulation
Glyceria acutiflora Torr.	2.53	0.38	6.66	+
Lolium perenne L.	1.15	0.58	1.98	+
Lophatherum gracile Brongn.	2.11	0.28	7.54	+
Phalaris arundinacea L. var. arundinacea	7.82	1.23	6.36	+
Pharagmites communis Trin.	1.54	0.38	4.05	+
Poa annua L.	4.26	0.90	4.73	+
Poa pratensis L.	2.31	0.56	4.13	+
Trisetum bifidum Ohwi.	2.70	0.74	3.65	+
Panicoideae				
Andropogen virginicus L.	1.48	0.38	3.89	+
Arthraxon hispidus Makino	1.90	0.62	3.06	+
Coix lacryma-jobi L.	5.84	1.25	4.67	+
Cymbopogon citratus Stapf	1.17	0.42	2.79	+
Digitaria ciliaris Koeler	4.34	1.05	4.13	+
D. radicosa Miq.	2.18	0.76	2.87	+
D. violascens Link	2.70	1.27	2.13	+
Eriochloa villosa Kunth	2.01	0.94	2.14	+
Hermarthria sibirica Ohwi	4.27	0.75	5.69	+
Imperata cylindrica Beauv.	1.28	0.47	2.72	+
Isachne globosa Kuntze	5.08	0.42	12.10	+
Microstegium nudum A. Camus	2.07	0.66	3.14	+
Microstegium vimineum A. Camus	3.02	0.86	3.51	+
Miscanthus sacchariflorus Benth. et. Hook. f.	1.42	0.35	4.05	+
Miscanthus sinensis Anderss.	3.20	0.91	3.52	+
Oplismenus undulatifolius Roem. et Schult.	3.10	0.80	3.88	+
Panicum bisulcatum Thunb.	2.16	1.70	1.27	+
Panicum crus-galli L.	4.12	1.26	3.27	+
P. crus-galli L. var. echinata Makino	3.64	1.49	2.44	+
P. crus-galli L. var. frumentaceum Trin.	1.43	1.11	1.29	+
Panicum dichotomiflorum Michx.	1.40	0.53	2.64	+
Paspalum distichum L.	3.89	0.67	5.81	+
Paspalum orbiculare Forst. f.	3.58	1.15	3.11	+
Paspalum thunbergii Kunth	4.57	0.87	5.25	+

	Si%	Ca%	Si/Ca	Degree of Si accumulation
Pennisetum japonicum Trin.	2.02	0.78	2.59	+
Saccharum spontaneum L. var. arenicola Ohwi	2.17	1.52	1.43	+
Setaria glauca Beauv.	2.51	0.96	2.61	+
Setaria viridis Beauv.	4.14	1.12	3.70	+
Sorghum halepense Per.	2.12	1.35	1.57	+
Themeda japonica C. Tanaka	1.00	0.71	1.41	+
Eragrostoideae				
Eleusine indica Gaertn.	1.68	2.00	0.84	±
Eragrostis cilianensis Link ex Lutati	2.66	0.97	2.74	+
Eragrostis curvula Nees	0.84	0.61	1.38	+
Eragrostis ferruginea Beauv.	2.55	0.64	3.98	+
Eragrostis multicaulis Steud.	1.35	0.79	1.71	+
Eragrostis pilosa Beauv.	1.87	0.81	2.31	+
Sporobolus fertilis W. Clayton	1.38	0.53	2.60	+
Zoysia japonica Steud.	1.49	0.39	3.82	+

Appendix 3-E

Content of Si and Ca in Commelinaceae collected from Kyoto Prefectural Botanical Garden

	Si %	Ca %	Si/Ca	Degree of Si accumulation
Commelinaceae				
Aneilema aequinoctiale Kunth	0.70	1.47	0.48	±
Ballya zebrine Brenan	1.03	1.62	0.64	±
Callisia elegans Alexander ex H.E.Moore	0.11	1.71	0.06	-
Callisia fragrans Woodson	1.39	2.57	0.54	±
Callisia multiflora H.E.Moore	0.98	1.66	0.59	±
Camperia zanonia H.B.K.	1.44	2.55	0.56	±
Commelina benghalensis L.	0.48	1.91	0.25	-
Commelina communis L.	1.52	1.15	1.32	+
Cyanotis somaliensis C.B.Clarke	0.16	2.13	0.08	-
Dichorisandra thyrsiflora Mikan	1.26	2.07	0.61	±
Gibasis geniculata	1.19	2.21	0.54	±
Rhoeo discolor Stearn	0.26	3.04	0.09	-
Setcreasea purpurea Boom	0.77	3.60	0.21	±
Sinderasis fuscata H.E.Moore	0.42	1.12	0.38	-
Tradescantia albiflora Kunth emend. Brucken	1.19	1.90	0.63	±
Tra. blossfeldiana Mildbr.	2.26	3.63	0.62	±
Tra.fluminensis Vell. emend. Brucken	1.69	2.32	0.73	±
Tra. holoserisea Jacq.	4.68	2.29	2.04	+
Tra. reflexa L.	0.18	1.82	0.10	-
Tra. sillamontana Matuda	0.71	3.78	0.19	±
Tra. subaspera Ker-Gawl.	0.17	1.56	0.11	-
Tra. venezuelensis Steyrm.	3.43	2.24	1.53	+
Tripogandra amplexicaulis Woods.	1.96	1.63	1.20	+
Zebrina focculosa Bruckn.	0.77	2.00	0.39	±
Z. pendula Schitzlein	0.85	4.38	0.19	±
Z. pendula Schitzlein [discolor]	1.80	3.01	0.60	±
Z. pendula Schitzlein [quadricolor]	1.32	2.70	0.49	±
Z. purpusii Bruckn. var. minor Rehn.	1.69	2.98	0.57	±
Average of 28 species	1.23	2.32	0.56	±

Appendix 3-F

Content of Si and Ca in Juncaceae collected from Hiroshima Prefectural Agricultural Experiment Station

	Si (%)	Ca(%)	Si/Ca	Degree of Si accumulation
Juncaceae				
Juncus effusus L. var. decipiens Buchen.*	0.14	0.24	0.58	-
J. setchuensis Buchen. var. effusoides**	0.15	0.24	0.64	-
* (♀) ×** (♂)	0.42	0.32	1.31	±
**(♀) ×* (♂)	0.60	0.35	1.72	±
J. effusus L. var. decipiens Buchen. [Sazanami]	0.29	0.28	1.07	±
[Sazanami 6.6mR]	0.38	0.33	1.13	±
[Isonami]	0.34	0.27	1.27	±
[Asanagi]	0.26	0.25	1.03	±
[Okayama-3]	0.34	0.26	1.32	±
[Komatsuzairai]	0.44	0.29	1.50	±
[Shimomasudazairai]	0.39	0.28	1.40	±
J. effusus L. f. spiralis	0.16	0.31	0.52	-
Avarage of 12 cultivars	0.33	0.29	1.12	±

Appendix 3-G

Content of Si and Ca in Cucurbitaceae collected from Experimental Farm of Okayama University

	Si (%)	Ca (%)	Si/Ca	Degree of Si accumulation
MAGNOLIOPHYTA (ANGIOSPERMAE)				
MAGNOLIOPSIDA (DICOTYLEDONEAE)				
Cucurbitaceae				
Citrullus vulgaris Schrad.	2.13	4.66	0.46	±
Cucumis melo L. var. makuwa Makino [Kikumelon]	1.96	4.42	0.44	±
Cucumis melo L. var. makuwa Makino [Ohgonsennari]	4.28	7.17	0.60	±
Cucumis sativus L.	1.53	4.35	0.35	±
Cucurbita moschata Poir. var. melonaeformis Makino	1.58	2.80	0.56	±
Lagenaria siceraria Standley var. hispida Hara	1.43	3.68	0.39	±
Luffa cylindrica Roem.	1.91	3.44	0.56	±
Momordica charantica L.	1.91	4.71	0.41	±
Average of 8 species	2.09	4.40	0.47	±

Appendix 3-H

Content of Si and Ca in Urticaceae collected from Kyoto Prefectural Botanical Garden

	Si (%)	Ca (%)	Si/Ca	Degree of Si accumulation
Urticaceae				
Pellionia radicans Wedd.	1.07	1.31	0.82	±
Pilea cadierei Gagnep et Guillaum	0.80	6.53	0.12	±
Pilea hamaoi Makino	1.22	6.58	0.19	±
Pilea petiolaris Blume	0.61	3.76	0.16	±
Pilea spruceana Wedd. [Silver-tree]	1.47	5.04	0.29	±
Average of 5 species	1.03	4.64	0.32	±

Appendix 3-I

Si accumulation in 4 sub-families of Gramineae

	Number of species	Si (%)	Ca (%)	Si/Ca
Bambusoideae	151	5.58	0.57	11.6
Oryzeae *	73	7.36	0.44	17.9
Others	78	3.91	0.69	5.81
Pooideae	22	2.69	0.63	4.62
Panicoideae	30	2.73	0.90	3.37
Eragrostoideae	8	1.73	0.84	2.45
Average of Gramineae	221	4.73	0.63	9.38

Site of sampling (except *): Kyoto Prefectural Botanical Garden
* National Institute of Genetics (Mishima)

Appendix 3-J

Distribution of Si accumulator in Pteridophyta

	Species number	Si (%)	Ca (%)	Si/Ca
Lycopsida				
*Selaginellaceae	2	4.60	0.55	8.42
Equisetapsida				
*Equisetaceae	2	5.81	2.18	2.79
Filicopsida				
*Marattiaceae	1	1.66	1.23	1.35
*Osumndaceae	2	4.01	1.04	4.17
*Blechnaceae	1	2.35	1.11	2.12
*Pteridaceae	4	2.67	0.84	3.05
*Thelypteridaceae	4	3.15	1.98	1.70
*Athyriaceae	7	1.86	1.52	1.20
**Davalliaceae	2	0.37	1.24	0.44
**Polypodiaceae	3	0.12	0.80	0.13
**Dryopteridaceae	17	0.27	1.42	0.18
Average of total	45	1.66	1.33	1.25
*Accumulator (A)	23	3.01	1.38	2.18
**Non-accumulator (B)	22	0.26	1.26	0.21
A/B		11.6	1.10	10.38

Site of sampling : Kyoto Prefectural Botanical Garden

Appendix 3-K

Water-soluble SiO_2 content in the soils of sampling sites

Sampling site	SiO_2 mg/100g soil
A	2.2
B	3.9
C	7.0
D	2.2
E	2.7

A : Nippon Shinyaku Institue for Botanical Research (Kyoto)
B : Kyoto Prefectural Botanical Garden
C : National Institute of Genetics (Mishima)
D : Hiroshima Prefectural Agricultural Experiment Station
E : Experiment Farm of Okayama University

Appendix 4-A

Silicon content of barley grain (Standard Variety, SV)

SV No.	Cultivar	Covered /naked[a]	Kernel row[b]	OU number[c]	Si content (mg/kg)
1	Compana (CI 5438)	+	2	A002	1786.6
2	Sanalta (CI 6087)	+	2	A008	2316.8
3	Olympia (CI 6107)	+	6	A321	2813.7
4	Vantage (CI 7324)	+	6	A604	2234.9
5	Peatland (CI 5267)	+	6	A607	2812.3
6	Libia	+	6	B008	2599.2
7	Rabat	+	6	B015	2524.2
8	Morocco	+	6	B033	2787.7
9	Giza 117	+	6	B069	1943.7
10	Macha	+	6	B318	2595.8
11	Tunis (CI 1383)	+	6	B327	2603.4
12	Giza 68	+	6	B349	1928.7
13	Cairo 1 (14)	+	6	B369	2141.8
14	Algiers	+	6	B623	3222.7
15	Suez (84)	+	6	B669	2372.9
16	Palmella Blue	+	2	B670	2812.5
17	Vladivostok	+	6	C001	2898.0
18	Pukou 1	+	6	C018	3234.1
19	Chiaochuang 1	+	6	C040	2226.6
20	Chengchou 3	+	6	C043	2346.0
21	Changchou 1	-	6	C045	1.0
22	Tibet Violet 1	-	6	C059	8.8
23	Manchuria Native 1	+	6	C302	2649.8
24	Sanchiang Fuchin	+	6	C304	2564.1
25	Chihchou	+	6	C319	2732.7
26	Titienchiao 2	+	6	C328	2625.6
27	Tayeh 1	-	6	C331	35.2
28	Paoanchen 1	+	6	C336	3252.8
29	Shanghai 1	-	6	C346	62.4
30	Tinghsien	-	6	C616	34.3
31	Wuhu	+	6	C619	3087.7
32	Takungkuan	-	6	C621	1.0
33	Tawangmiao 1	+	6	C624	2830.4
34	Mushinchiang 3	+	6	C627	2741.9
35	Tibet White 4	-	6	C656	5.4
36	Debre Zeit 1 (1-5-17a)	+	6	E216	2387.5

SV No.	Cultivar	Covered /naked[a]	Kernel row[b]	OU number[c]	Si content (mg/kg)
37	Addis Ababa 40 (12-24-84)	+	L	E245	2269.3
38	Sululta 1 (1-8-11)	+	6	E254	3434.3
39	Mota 1 (1-24-10)	+	D	E269	2751.9
40	Dabat 1 (1-27-25)	-	6	E278	22.2
41	Molale 1 (2-8-4a)	+	L	E284	2108.6
42	Debra Birhan 1 (2-8-18)	+	6	E285	2265.4
43	Debre Zeit 29 (1-5-35)	+	D	E525	1966.4
44	Addis Ababa 56 (3-10-1b)	-	6	E550	20.2
45	Nazareth 1 (1-10-1a)	+	D	E559	2918.6
46	Asbe Tafari 1 (1-11-19)	+	6	E560	2962.2
47	Gondar 1 (1-27-1)	+	6	E574	2424.5
48	Quiha 1 (2-3-73)	+	D	E581	2219.1
49	Dembi 1 (2-20-51)	+	L	E589	2729.8
50	Debre Zeit 18 (1-5-28a)	+	L	E821	2606.0
51	Addis Ababa 3 (12-24-9)	+	6	E832	2269.4
52	Jijiga 2 (1-13-20)	+	6	E862	2315.0
53	Kulubi 3 (1-14-38b)	+	L	E863	2269.2
54	Deder 1 (1-16-27)	+	L	E864	2649.2
55	Glyorgi 1 (1-24-21a)	+	D	E872	1985.1
56	Adi Abun 1 (1-28-2)	+	D	E879	2107.8
57	Dessie 1 (2-4-22)	+	6	E883	2388.1
58	Ammora 1 (99-1)	+	6	F001	2790.5
59	Zeggi 1 (137-1)	+	D	F003	2353.7
60	Quesi (471)	+	D	F035	3033.8
61	Palagi (499)	+	6	F037	2694.1
62	Adigrat 3 (508-2)	+	L	F038	2469.0
63	Fiche 1 (212a-1)	+	L	F313	2085.7
64	Sheki 2 (260-2)	+	6	F318	2958.5
65	Sombo 1 (281-1)	+	L	F322	2500.8
66	Asella 1 (372a-1)	+	D	F324	3629.4
67	Mai Chew 1 (495)	+	L	F336	3370.6
68	Debra Markos 1 (181-1)	+	L	F606	2985.3
69	Jimma 2 (265-1)	+	6	F619	2455.2
70	Shashamane 1 (415)	+	6	F629	2857.2
71	Meki 2 (481-2)	+	L	F635	2198.5
72	Adigrat 8 (520c-2)	+	L	F639	2675.6

SV No.	Cultivar	Covered /naked[a]	Kernel row[b]	OU Number[c]	Si content (mg/kg)
73	Ballia	+	6	I003	2435.2
74	Kabul 1	+	6	I012	1711.3
75	Iraq Black Barley	+	2	I024	3084.0
76	Aleppo 1 (438)	+	2	I027	2903.4
77	Esfahan 1 (152 III)	+	6	I032	2007.6
78	Gorgan 1 (225A I)	+	6	I036	2664.3
79	Khanaqin 2 (KUH5305)	+	2	I166	3019.3
80	Khanaqin 5 (KUH5308)	+	2	I167	2704.8
81	Arbat 1 (KUH 5317)	+	2	I170	2553.9
82	Chamchamal 1 (KUH 5326)	+	2	I173	2684.8
83	Samarra 1 (KUH 5329)	+	6	I174	2418.6
84	Karand 1 (KUH 5365)	+	2	I182	2452.5
85	Rewari	+	6	I304	1562.4
86	Charikar 2 (606 II)	-	6	I323	13.0
87	Iraq Barley 1	+	6	I325	1943.4
88	Katana 2 (183)	+	2	I326	3127.0
89	Aleppo 2 (439)	+	2	I327	2284.8
90	Damaneh 1 (165d)	+	6	I334	1711.1
91	Ghazvin 1 (184)	+	2	I335	1784.6
92	Firuzkuh 1 (277)	+	6	I337	1240.1
93	Sarab 1 (347 I)	+	2	I339	1916.2
94	Dunga Banse (N 5)	-	6	I341	3.8
95	Happar 1 (N 14)	-	L	I343	10.7
96	Baiji 2 (KUH 5333)	+	2	I475	2290.3
97	Karad	+	6	I607	2621.2
98	Quetta 1 (14-1)	-	6	I617	68.6
99	Chaman 1 (47 II)	+	6	I618	2184.1
100	H.E.S. 4 (Type 12)	+	6	I622	2020.2
101	Katana 1 (182)	+	2	I626	3331.9
102	Ardabil 1 (336 II)	+	6	I639	2326.7
103	Baqubah 1 (KUH 5301)	+	6	I764	1886.6
104	Khanaqin 1 (KUH 5304)	+	2	I765	2753.9
105	Khanaqin 4 (KUH 5307)	+	2	I766	2600.0
106	Sulaymaniyah 2 (KUH 5322)	+	2	I771	2707.6
107	Sinjar 1 (KUH 5337)	+	2	I776	2529.8
108	Gozen	+	6	J016	2564.5
109	Hanbozu	+	6	J040	2900.7

SV No.	Cultivar	Covered /naked[a]	Kernel row[b]	OU Number[c]	Si content (mg/kg)
110	Hosomugi	+	6	J044	3406.8
111	Hayakiso 2	+	6	J064	2561.2
112	Yahazu	-	6	J069	29.1
113	Shishikui Zairai	-	6	J075	26.6
114	Irino Zairai	+	6	J080	2582.0
115	Shimabara	-	6	J087	0.0
116	Sazanshu	+	6	J097	2655.6
117	Yatomi Mochi	-	6	J183	10.8
118	Sekitori	+	6	J326	2946.4
119	Suifu	+	6	J330	2994.2
120	Honen	-	6	J334	5.4
121	Chikurin	+	6	J338	2910.5
122	Wase Bozu	+	6	J366	2540.9
123	Kobinkatagi	-	6	J369	38.1
124	Hachikoku	+	6	J386	2858.4
125	Tainan	+	6	J395	2580.8
126	Zairai 1	+	6	J396	2530.4
127	Bozu Mochi	-	6	J467	23.9
128	Shirochinko	-	6	J639	0.0
129	Akashinriki	-	6	J647	0.3
130	Shiroto	-	6	J669	0.0
131	Shikke Shirazu	-	6	J672	0.3
132	Wase Hadaka	-	6	J674	20.7
133	Hizahachi	-	6	J682	112.2
134	Hitokawa	+	6	J689	2183.8
135	Hayatori Hadaka	-	6	J691	36.2
136	Zairai 2	+	6	J696	2086.9
137	Taihoku A	+	6	J697	2014.2
138	Goheung Covered 1	+	6	K001	2253.1
139	Goheung Naked 2	-	6	K002	23.5
140	Yeongam Covered 2	+	6	K003	2291.0
141	Masan Naked 1	-	6	K043	30.7
142	Waegwan Covered 1	+	6	K059	2372.4
143	Yeongcheon Covered 2	+	6	K060	1586.1
144	Anseong Covered 2	+	6	K094	1498.2
145	Namyang Covered 2	+	6	K095	2229.7
146	Gangneung Covered 3	+	6	K108	2626.8
147	Hamjong Covered 1	+	6	K113	2680.8
148	Hongweon Inuno-o	+	6	K117	2644.4

SV No.	Cultivar	Covered /naked[a]	Kernel row[b]	OU Number[c]	Si content (mg/kg)
149	Boseong Covered 3	+	6	K304	2202.1
150	Gwangju Covered 4	+	6	K311	1501.8
151	Buyong Oni Hadaka 2	-	6	K327	39.0
152	Dongsan Oni Hadaka 2	-	6	K329	0.0
153	Masan Covered 5	+	6	K342	1987.8
154	Daecheon Covered 4	+	6	K374	2040.4
155	Yesan Covered 2	+	6	K377	2111.8
156	Baegcheon Covered 1	+	6	K401	2310.2
157	Gyeongseong Rokkaku	+	6	K419	1938.1
158	Jangheung Naked 2	-	6	K602	28.2
159	Sunchang Naked 4	-	6	K624	15.0
160	Mangyeong Naked 3	-	6	K625	70.4
161	Tongyeong Covered 1	+	6	K638	2100.8
162	Gyeongju Covered 4	+	6	K655	2672.8
163	Janghang Covered	+	6	K672	2779.6
164	Yeongdong Seungmaeg 1	+	6	K689	2694.0
165	Eumseong Covered 3	+	6	K692	2152.4
166	Namcheon Covered	+	6	K702	1734.5
167	Hongcheon Covered 1	+	6	K706	2192.4
168	Sama 1 (1385)	-	6	N005	34.1
169	Tilman Camp 1 (1398)	-	6	N009	30.8
170	Pisang 1 (1427)	+	6	N016	2063.3
171	Thangja 1 (1433)	-	6	N017	32.1
172	Birkna Camp 1 (1487)	+	6	N031	2339.7
173	Kagbeni 1 (1616)	+	6	N051	2036.0
174	Ulleri 1 (1633)	+	6	N055	2030.4
175	Sipche 1	-	6	N077	46.4
176	Gho 1 (1392)	-	6	N308	70.8
177	Katmandu 2 (1438)	+	6	N319	2833.1
178	Chame 1 (1493)	+	6	N333	3113.2
179	Ngyak 1 (1524)	-	6	N340	30.8
180	Prok 1 (1537)	-	6	N344	56.9
181	Lumley 1 (1674)	+	6	N369	2304.7
182	Annapurna B.C. 1 (1422)	-	6	N615	33.4
183	Trisuli Bazar 3 (1449)	+	6	N622	3489.6
184	Kakani Bangalow 3 (1477)	+	6	N628	2937.2
185	Macha Khola 1 (1483)	+	6	N630	2806.4
186	Dhumpu 1 (1596)	+	6	N647	2180.8

SV No.	Cultivar	Covered /naked[a]	Kernel row[b]	OU number[c]	Si content (mg/kg)
187	Sikha 2 (1626)	-	6	N653	34.4
188	Keronja 1 (1651)	+	6	N661	2705.9
189	Turkey 1	+	6	T001	2212.7
190	Turkey 31	+	2	T011	2242.6
191	Turkey 61	+	2	T021	2525.7
192	Bursa (918)	+	6	T267	2684.5
193	Amasya (1208)	+	2	T268	2230.2
194	Turkey 11	+	6	T304	1899.6
195	Turkey 41	+	2	T314	1924.0
196	Turkey 71	+	6	T324	2204.0
197	Turkey 101	+	6	T334	1917.5
198	Rerhauli 1 (454)	+	6	T566	2922.4
199	Goenen (997)	+	6	T567	3090.5
200	Ayas (1309)	+	2	T568	2754.5
201	Turkey 21	+	6	T607	2558.5
202	Turkey 51	+	2	T617	2360.4
203	Turkey 81	+	2	T627	2496.3
204	Turkey 111	+	2	T637	1943.0
205	Istanbul (1024)	+	2	T867	2008.8
206	Ankara (1369)	+	2	T868	2129.2
207	Samaria 4 Zeilige	+	6	U005	2733.6
208	Bolognes	+	6	U009	2457.3
209	Orayio	+	6	U013	2276.0
210	Bulgarian 447	+	6	U015	2506.7
211	Kolnozni	+	6	U019	2535.8
212	Darmatishe	+	6	U023	2749.1
213	Balkan 2	+	6	U025	3492.1
214	Rene	+	2	U029	2138.5
215	Kleinwanz	+	6	U030	2481.4
216	Adliker	+	2	U046	3107.8
217	Michalovicky nahy	-	2	U048	21.1
218	Jubilee	+	2	U049	2543.9
219	Prentice	+	2	U050	1972.0
220	Archer	+	2	U051	1942.1
221	Maja	+	2	U053	2121.0
222	Erhart Frederickson	+	6	U055	2254.2
223	Tammi	+	6	U059	2675.3
224	Baku 1 (Cauc. 1)	+	6	U089	3367.0
225	Erevan 1 (Cauc.15)	+	2	U090	2028.6

SV No.	Cultivar	Covered /naked[a]	Kernel row[b]	OU number[c]	Si content (ppm)
226	Shemakha 1 (Cauc.37)	+	6	U094	1803.0
227	Erevan 6 (Cauc.40)	+	2	U095	2315.2
228	CSR 82	+	2	U145	2539.6
229	KUH 837	+	2	U172	2458.5
230	Badajoy	+	6	U305	2369.6
231	Caveda	+	6	U316	2562.4
232	Rumanian 19	+	6	U317	2279.0
233	Urania	+	6	U322	2462.3
234	Hungarian	+	D	U325	2589.0
235	France 1	+	2	U326	1739.1
236	Albert	+	6	U327	1811.0
237	Cygne	+	2	U329	2110.3
238	Oldenburger Landgerste	+	6	U340	2212.7
239	Saalegerste	+	2	U341	1980.0
240	Dometzkoer Paradies	-	2	U347	43.2
241	Plumage	+	2	U351	1742.4
242	Imperial	+	2	U352	2759.2
243	Opal	+	2	U353	1819.0
244	Drost	+	2	U354	1720.2
245	Vega	+	6	U355	2640.9
246	Olli	+	6	U359	2332.4
247	Caucasus	+	2	U388	2491.5
248	Geghard 1 (Cauc.44)	+	2	U397	2404.3
249	Gori (Cauc.53)	+	6	U400	3142.5
250	PLD 5	+	2	U434	1590.1
251	CSR 3	+	2	U441	1990.0
252	Ribofardo	+	6	U603	2063.5
253	Valencia	+	6	U607	2548.3
254	Tripoli	+	6	U610	2152.8
255	Otello	+	6	U613	2448.1
256	Bulgarian 347	+	6	U615	3249.1
257	Moldavia	+	6	U616	2383.6
258	Aurore	+	2	U628	2174.7
259	Hanna	+	2	U633	2243.0
260	Bavarian	+	2	U640	2665.0
261	Bohemian	+	2	U641	2122.7
262	Tivannes	+	2	U646	2076.7
263	Prokupkuy nahy	-	2	U647	9.2
264	Tuxsky nahy	-	6	U648	20.5

SV No.	Cultivar	Covered /naked[a]	Kernel row[b]	OU number[c]	Si content (ppm)
265	Binder	+	2	U652	2278.6
266	Ymer	+	2	U658	2054.9
267	Vankhuri	+	2	U659	1927.4
268	Tibilisi 1 (Cauc.20)	+	6	U692	3168.4
269	Baku 6 (Cauc.35)	+	6	U694	2305.0
270	PLD 49	+	2	U737	2074.3
271	PLD 83c	+	2	U738	2417.7
272	PLD 139	+	2	U740	1733.8
273	KUH 836	+	6	U771	3542.3
274	KUH 842	+	2	U773	2855.6

[a] -, Naked; +, Covered ; [b] 2, two-rowed; 6, six-rowed; D, deficiency; L, labile or irregulare; [c] OU, Okayama University (J, Japan; K, Korea; C, China; N, Nepal; I, South-west Asia including India; T, Turkey; U, Europe; B, North Africa; E, Ethiopia, A: America and others).

Thanks are given to Prof. Takeda at Research Institute for Bioresources, Okayama University for providing barley seeds.

Appendix 4-B

Silicon content of barley grain (Barley Core Collection of United State, BCCUS)

BCUS No.	Variety	Covered/ naked[a]	Kernel row[b]	Country	Si content (mg/kg)
1	AC OXBOW	+	2	Canada	3138.7
2	ACORA	-	6	Peru	102.9
3	ACUARIO	+	2	Chile	2191.6
4	ADVANCE	+	6	United States	3347.1
5	ALBANY	+	2	Canada	2353.1
6	ALISO	+	6	ICARDA-CIMMYT	2975.0
7	ANCAP 2	+	6	Uruguay	2956.0
8	ANTARCTICA 05	+	2	Brazil	2111.7
9	ANTARCTICA 06	+	2	Brazil	2681.3
10	APIZACO	+	6	Mexico	2812.1
11	ARIVAT	+	6	United States	3172.8
12	ARUPO	+	2	ICARDA-CIMMYT	2892.9
13	ATAHUALPA	-	2	Ecuador	234.6
14	ATLAS 68	+	6	United States	2792.8
15	BEACON	+	6	United States	2873.5
16	BEDFORD	+	6	Canada	3125.6
17	BEECHER	+	6	United States	2279.7
18	BELFORD	+	6	United States	2664.2
19	BETZES	+	2	Germany	2421.6
20	BOLIVIANA	+	3	Ecuador	2311.7
21	BONANZA	+	6	Canada	2804.0
22	BONNEVILLE 70	+	6	United States	NA
23	BOWMAN	+	2	United States	2771.1
24	BOYER	+	6	United States	2026.6
25	CABANILLAS	-	6	Peru	95.4
26	CALICUCHIMA	+	6	Ecuador	2880.4
27	CALIFORNIA COAST	+	6	United States	1579.8
28	CALIFORNIA MARIOUT	+	6	United States	2148.6

BCUS No.	Variety	Covered/ naked[a]	Kernel row[b]	Country	Si content (mg/kg)
29	CANADIAN THORPE	+	2	United Kingdom	2159.5
30	CAPACHICA	-	6	Peru	114.8
31	CAPACHICA 2	-	6	Peru	133.1
32	CAPACHICA 3	-	6	Peru	169.7
33	CASBON	+	6	United States	NA
34	CASCARUDA	+	6	Bolivia	2919.8
35	CDC BUCK	-	6	Canada	119.2
36	CDC RICHARD	-	2	Canada	NA
37	CENTAURO	-	2	Chile	165.0
38	CENTENNIAL	+	2	Canada	2089.4
39	CENTINELA	+	6	Mexico	2780.2
40	CERRO PRIETO	+	6	Mexico	3457.5
41	CERVECERA 3278	+	2	Uruguay	2102.9
42	CHAPAIS	+	6	Canada	2510.4
43	CHARLOTTETOWN 80	+	2	Canada	1852.5
44	CHIA	+	6	Colombia	2001.0
45	CHULLUNQUIANI	-	6	Peru	104.4
46	CM 67	+	6	United States	2801.2
47	COMPANA	+	2	United States	1722.9
48	COMUN	+	6	Mexico	NA
49	CONDOR	-	2	Canada	103.7
50	CONQUEST	+	6	Canada	2783.8
51	DAWN	+	6	United States	2501.2
52	DIAMOND	+	6	Canada	2917.7
53	DICKTOO	+	6	United States	2925.7
54	DORADA	+	6	Ecuador	3033.5
55	DUCHICELA	+	6	Ecuador	2638.9
56	ESMERALDA	+	6	Mexico	2288.0
57	ESPERANZA	+	6	Mexico	2766.7
58	FALCON	-	6	Canada	NA
59	FERGUS	+	2	Canada	2059.3
60	FNC 1	+	2	Uruguay	2205.7
61	FNC 6-1	+	2	Uruguay	2362.8
62	FORRAJERA 75	+	6	Uruguay	2634.1
63	FOSTER	+	6	United States	2746.4

BCUS No.	Variety	Covered/ naked[a]	Kernel row[b]	Country	Si content (mg/kg)
64	FRANCISCANA	+	6	Ecuador	2435.6
65	FRONTERA	+	6	Chile	NA
66	GALERAS	+	6	Colombia	2975.4
67	GALT	+	6	Canada	2771.8
68	GLORIA/COME	+	6	ICARDA-CIM MYT	2668.5
69	GOBERNADORA	+	2	ICARDA-CIMMYT	2574.9
70	GODIVA	-	6	United States	116.9
71	GRANIFEN	+	2	Chile	NA
72	GRANO COMUN	+	6	Bolivia	2652.4
73	HANNCHEN	+	2	Sweden	NA
74	HARRINGTON	+	2	Canada	2002.0
75	HAZEN	+	6	United States	2246.3
76	HECTOR	+	2	Canada	NA
77	HIMALAYA	-	6	United States	86.0
78	HORSFORD	+	6	United States	2569.3
79	HUCHUYMUJU	+	6	Bolivia	2593.2
80	HUDSON	+	6	United States	2313.6
81	HUSKY	+	6	Canada	NA
82	IBTA 80	+	6	Bolivia	2286.2
83	ILAVE	-	6	Peru	NA
84	JACKSON	+	6	United States	2817.5
85	KAMIAK	+	6	United States	2892.4
86	KARL	+	6	United States	2862.4
87	KEARNEY	+	6	United States	2149.7
88	KEYSTONE	+	6	Canada	2787.1
89	KINDRED	+	6	United States	2989.8
90	KINKORA	+	6	Canada	3346.3
91	KLAGES	+	2	United States	2614.4
92	KOLLA	+	6	Bolivia	2661.6
93	KOMBYNE	+	6	United States	1692.7
94	LARKER	+	6	United States	NA
95	LAUFEN	-	2	Chile	148.2
96	LECERVECERA 2034	+	2	Uruguay	2184.4
97	LEGER	+	6	Canada	2762.9
98	LEO	+	2	Chile	1749.9

BCUS No.	Variety	Covered/ naked[a]	Kernel row[b]	Country	Si content (mg/kg)
99	LIBRA	+	6	Chile	2018.9
100	LINO	-	6	ICARDA-CIMMYT	130.4
101	MANAPATA	-	6	Peru	148.7
102	MANCHURIA	+	6	United States	3845.7
103	MANLEY	+	2	Canada	3380.7
104	MONTCALM	+	6	Canada	3034.7
105	MORAVIAN	+	2	United States	2541.6
106	MOREX	+	6	United States	3079.3
107	MUNSING	+	2	United States	1655.0
108	OAC 21	+	6	Canada	3477.0
109	OAC ELMIRA	+	6	Canada	2836.5
110	ODESSA	+	6	United States	2345.3
111	OLLI	+	6	Finland	2463.7
112	OTAL	+	6	United States	2923.5
113	OTIS	+	2	United States	2252.3
114	PALERMO	-	6	Peru	99.5
115	PALLISER	+	2	Canada	2646.6
116	PEATLAND	+	6	United States	3293.5
117	PIROLINE	+	2	Germany	2224.2
118	POST	+	6	United States	2529.4
119	PRIMUS II	+	6	United States	3200.5
120	PUEBLA	+	6	Mexico	2793.4
121	QUIBENRAS	+	6	Colombia	2539.8
122	REX	+	2	Canada	2380.9
123	RIDAWN	+	2	United States	NA
124	RITA PELADA	-	6	Ecuador	60.4
125	ROBUST	+	6	United States	3383.3
126	RUSSELL	+	6	United States	3645.2
127	SACASCO 1	+	6	Peru	2372.2
128	SAN BENITO	-	6	Bolivia	149.3
129	SANALTA	+	2	Canada	NA
130	SCHUYLER	+	6	United States	2267.0
131	SHONKIN	-	2	United States	37.2
132	SHYRI	+	2	Ecuador	2792.4
133	STANDER	+	6	United States	2470.9
134	STEPTOE	+	6	United States	NA
135	SUAMOX	+	6	Colombia	3201.1
136	TAMBAR 500	+	6	United States	2465.4

BCUS No.	Variety	Covered/ naked[a]	Kernel row[b]	Country	Si content (mg/kg)
137	TARACO	-	6	Peru	121.3
138	TENNESSEE WINTER	+	6	United States	2445.9
139	TERAN	+	3	Ecuador	3554.0
140	TR 306	+	2	Canada	2167.1
141	TREBI	+	6	United States	2717.9
142	UNITAN	+	6	United States	NA
143	WARRIOR	+	6	Canada	3587.3
144	WEAL	+	6	United States	3787.3
145	WINTER CLUB	+	6	United States	NA
146	WISCONSIN PEDIGREE 38	+	6	United States	2959.7
147	WONG	+	6	China	2506.5
148	WYSOR	+	6	United States	2513.8
149	YANALA	+	6	Colombia	2749.1
150	YANICO	-	6	Peru	43.1
151	YORK	+	6	Canada	2204.9
152	YUNGUYO L3	-	6	Peru	68.0

[a] -, Naked; +, Covered ; [b] 2, two-rowed; 6, six-rowed

NA: not analysed.

Appendix 4-C

Silicon content of barley grain (Barley Core Collection of East Asia, BCCEA)

BCEA No.	Variety	Covered/ naked	Kernel row	Country	Si content (mg/kg)
1	TKB64	-	6	Bhutan	346.6
2	TKB69b	-	6	Bhutan	393.5
3	TKB73a	-	6	Bhutan	208.6
4	TKB73b	-	6	Bhutan	204.1
5	TKB75a	-	6	Bhutan	267.5
6	TKB75c	-	6	Bhutan	138.1
7	TKB79a	-	6	Bhutan	291.9
8	TKB80a	-	6	Bhutan	326.9
9	TKB80c	-	6	Bhutan	344.1
10	TKB81a	-	6	Bhutan	293.7
11	TKB81c	-	6	Bhutan	295.3
12	TKB82b	-	6	Bhutan	266.2
13	TKB82d	-	6	Bhutan	338.9
14	TKB82f	-	6	Bhutan	299.2
15	TKB83b	-	6	Bhutan	136.7
16	Dong Ning PI 7 Hao-1	+	6	China	3188.6
17	Hu Lan PI 1-2	+	6	China	2989.8
18	Bai Quan PI 7 Hao	+	6	China	2733.4
19	La Lin Luo 1 Hao	-	6	China	172.7
20	Ning Cheng Da Mai(1)	-	6	China	354.0
21	Xin Mai 1 Hao	+	6	China	2660.8
22	Jiang Ning 1281	+	6	China	3596.9
23	Gang Tuo Qing Ke 1 hao	-	6	China	345.4
24	Tu Da Mai 1	+	6	China	2558.2
25	Hu Mai 4 Hao	+	2	China	2863.2
26	Jing Pi 198	+	6	China	3748.2

BCEA No.	Variety	Covered/ Hulless	Spike row	Country	Si content (mg/kg)
27	E Dong 85-1	+	2	China	3312.4
28	Jing Ke 1 Hao	-	6	China	540.7
29	Jing pi C627-6	+	6	China	4007.0
30	Jing luo 2 hao	-	6	China	480.2
31	Han 85-222	+	2	China	2777.6
32	Fu 8	+	2	China	2457.5
33	Hu mai 10 hao	+	2	China	3084.1
34	E nong 82-6003	+	2	China	3382.7
35	E jing 145	+	2	China	2755.0
36	Vladivostok	+	6	China	2669.3
37	Harbin 13-8A	+	6	China	2665.8
38	Manchuria Native 1	+	6	China	2910.0
39	Fengtien Black	+	6	China	2158.2
40	Pinchiang Chaotung	+	6	China	2676.0
41	Sanchiang Fuchin	+	6	China	NA
42	Chientao Lungching	+	6	China	2853.1
43	Mongolia 6-row	-	6	China	191.8
44	Sanho	+	6	China	2731.2
45	Fuchin	+	6	China	2725.9
46	Harbin Native	+	6	China	NA
47	Mulan	+	6	China	2855.2
48	Taonan	+	6	China	2807.7
49	Tungfeng	+	6	China	2406.9
50	Manchuria 1	+	6	China	2598.8
51	Lutai	+	6	China	2388.6
52	Hsin Hsien	-	6	China	131.3
53	Shantung Naked	-	6	China	98.9
54	Litsun 2	+	6	China	1958.1
55	Chiaohsien 5	+	6	China	2133.4
56	Changtien 1	+	6	China	1819.1

BCEA No.	Variety	Covered/ Hulless	Spike row	Country	Si content (mg/kg)
57	Pukou 2	+	6	China	2545.7
58	Wuhu	+	6	China	2530.9
59	Tatung	+	6	China	2217.1
60	Chihchou	+	6	China	2765.4
61	Takungkuan	-	6	China	251.0
62	Tungliu	+	6	China	2494.2
63	Pantse 1	-	6	China	198.0
64	Liussuchiao 1	+	6	China	3113.2
65	Tawangmiao 1	+	6	China	3249.0
66	Chiuchiang	+	6	China	2826.7
67	Juichang 1	+	6	China	3100.6
68	Mushinchiang 3	+	6	China	3578.5
69	Yanghsin 3	+	6	China	2554.6
70	Titienchiao 2	+	6	China	2239.6
71	Paishapu 2	-	6	China	218.8
72	Paisha Tayeh 1	+	6	China	3360.3
73	Tayeh 4	-	6	China	220.0
74	Tayeh 11	+	6	China	3440.9
75	Paoanchen 1	+	6	China	3707.9
76	Chinniu 1	-	6	China	229.4
77	Chinniu 3	+	6	China	3173.7
78	Hsin-antien 1	+	6	China	2048.4
79	Chiaochuang 6	+	6	China	2776.0
80	Chengchou 5	+	6	China	2776.4
81	Changchou 1	-	6	China	222.7
82	Suchou 1	-	6	China	NA
83	Shanghai 1	-	6	China	214.5
84	Shanghai 3	-	6	China	249.1
85	Shanghai 7	-	6	China	233.7
86	Tohoku Shiro Hadaka	-	6	China	288.3
87	Violaceum Sg Type 3	-	6	China	418.7

BCEA No.	Variety	Covered/ Hulless	Spike row	Country	Si content (mg/kg)
88	Coeleste Se Type 4	-	6	China	317.3
89	Himalayense Type 5	-	6	China	279.5
90	Tibetanum Sh Type 6	-	6	China	264.2
91	Hangaicum Type 8	-	6	China	472.5
92	Violaceum Sh Type 10	-	6	China	303.3
93	Revelatum Type 13	-	6	China	272.3
94	Asiaticum Type 14	-	6	China	207.6
95	Sikangense Type 15	-	6	China	343.9
96	Kobdicum Type 20	-	6	China	851.9
97	Tibet White 9	-	6	China	164.9
98	Tibet White 16	-	6	China	196.4
99	Tibet White 25	-	6	China	194.3
100	Tibet Violet 1	-	6	China	226.9
101	Violaceum 2 (China)	-	6	China	559.0
102	Itu Native	+	6	China	3800.1
103	Guang da mai	-	6	China	203.6
104	Ying chun da mai	-	6	China	466.4
105	Luo ren da mai	-	6	China	1107.1
106	Quan zhi da mai	+	6	China	1714.4
107	Hai lun pi 4 hao	+	6	China	3605.6
108	Ke shan da mai	+	6	China	3401.7
109	Ke shan xi cheng da mai	+	6	China	2718.8

BCEA No.	Variety	Covered/ Hulless	Spike row	Country	
110	Ding xi da mai	+	6	China	3245.8
111	Zi gan liu leng	+	6	China	2348.1
112	Chuan sha wan dai mai	+	6	China	2867.9
113	Zhong da mai	+	6	China	3337.2
114	Liu leng zi da mai	+	6	China	3369.1
115	Lao wu hu xu mai	+	2	China	2672.1
116	Wu shen yang cao mai	+	2	China	2766.1
117	Wu shen da mai	+	6	China	3996.9
118	Dong sheng da mai	+	6	China	2091.7
119	Yuan mai	-	6	China	1408.7
120	Mii mai	-	6	China	940.7
121	Bilara-2	+	6	China	3482.1
122	Azad	+	6	China	4359.1
123	DL 88	+	6	China	2580.5
124	Rajkiran	+	6	India	3297.5
125	RD 2052	+	6	India	3077.2
126	C-138	+	6	India	3013.1
127	BHS169	+	6	India	2584.3
128	Sonu	+	6	India	2467.0
129	VLB-1	+	6	India	2941.8
130	HBL 113	+	2	India	1748.2
131	Dolma	-	6	India	0.0
132	Ratna	+	6	India	1825.0
133	Lakhan	+	6	India	2048.5
134	Karan 16	-	6	India	0
135	PL 172	+	6	India	2577.6
136	K. 12	+	6	India	2422.6
137	Ballia	+	6	India	1872.0
138	Rewari	+	6	India	1376.5

BCEA No.	Variety	Covered/ Hulless	Spike row	Country	Si content (mg/kg)
139	Patan	+	6	India	1762.8
140	Skoro	+	6	India	1710.4
141	Minapin	-	6	India	34.41
142	Milgagar	-	6	India	NA
143	Pindras 2	-	L	India	0
144	Dras 2	-	6	India	0
145	Pharona	-	6	India	0
146	Yagza 3	-	6	India	0
147	Partek 2	-	6	India	0
148	Skianku	-	L	India	0
149	Shargundik 2	-	6	India	81.2
150	Phokar 2	-	L	India	107.7
151	Churtanchan 1	+	6	India	1479.0
152	Chispiyanzan	-	L	India	467.3
153	Kangi 2	-	L	India	217.9
154	Khalsi 1	-	L	India	252.0
155	Narula 2	-	6	India	470.7
156	Saspul 2	-	6	India	336.3
157	Shakti 1	-	6	India	212.5
158	Martselang 2	-	6	India	395.7
159	Gat	+	6	India	2908.0
160	Tibba 1	+	6	India	2818.5
161	Sangrambata	+	6	India	2782.5
162	Dul	+	6	India	2919.8
163	Hanswani	+	6	India	3063.5
164	Kela	+	6	India	3275.0
165	Salooni 2	+	6	India	2191.9
166	Batheri	+	6	India	3385.5
167	Bag	+	6	India	3595.1
168	Nammu	-	6	India	393.7
169	Kawaring 2	-	6	India	341.8
170	Jari 1	+	6	India	2923.1
171	Sangri	+	6	India	3001.1
172	Surnji	+	6	India	2754.0

BCEA No.	Variety	Covered/ Hulless	Spike row	Country	Si content (mg/kg)
173	Chaundhar	+	6	India	2318.7
174	Dhaneti 2	-	6	India	373.4
175	Pholad Santarn 1	+	6	India	2570.4
176	Jamu 1	+	6	India	2863.5
177	Pandukeshwar 2	+	6	India	2936.4
178	Tefna	-	6	India	478.1
179	Bijoria	+	6	India	2724.7
180	Kunjanpur 1	+	6	India	2433.3
181	Bana	+	6	India	2194.0
182	Mansinghkanda 4	+	6	India	2112.4
183	Dhun 1	+	6	India	2762.5
184	Gauriat 1	+	6	India	3646.0
185	Timladigi 2	+	6	India	4003.5
186	Azumamugi	+	6	Japan	2619.1
187	Hoteimugi	+	6	Japan	3025.6
188	Asahi 19	+	2	Japan	2305.2
189	Satsuki Nijo	+	2	Japan	1798.7
190	Fuji Nijo	+	2	Japan	2377.1
191	Hinode Hadaka	-	6	Japan	186.0
192	Kikai Hadaka	-	6	Japan	169.2
193	Nanpu Hadaka	-	6	Japan	163.5
194	Hoshimasari	+	2	Japan	2537.8
195	Benkeimugi	+	6	Japan	2789.9
196	Minorimugi	+	6	Japan	3077.1
197	New Golden	+	2	Japan	2498.5
198	Azuma Golden	+	2	Japan	2173.7
199	Daisen Gold	+	2	Japan	2442.9
200	Shiratama Hadaka	-	6	Japan	123.9
201	Amagi Nijo 3	+	2	Japan	1994.2
202	Senbon Hadaka	-	6	Japan	0.0
203	Ishuku Shirazu	+	2	Japan	3382.9

BCEA No.	Variety	Covered/ Hulless	Spike row	Country	Si content (mg/kg)
204	Kawamizuki	+	2	Japan	2836.1
205	Haruna Nijo	+	2	Japan	2338.1
206	Misato Golden	+	2	Japan	2533.1
207	Sanadamugi	+	6	Japan	3378.6
208	Asamamugi	+	6	Japan	3114.0
209	Sanuki Hadaka	-	6	Japan	0.0
210	Hayate Hadaka	-	6	Japan	54.5

REFERENCES*

Ahmad, R., Zaheer, S.H. and Ismail, S. 1992 Role of slicon in salt tolerance of wheat (*Triticum aestivum* L.). Plant Sci. 85:43-50.

Agarie, S., Agata, W., Uchida, H., Kubota, F. and Kaufman, P. B. 1996. Function of silica bodies in the epidermal system of rice (*Oryza sativa* L.): testing the window hypothesis. J. Exp. Bot. 47: 655-660.

Ando, J. and Nakajima, J. 1985. Mineral composition and solubility of potassium silicate fertilizer. Jpn. J. Soil Sci. Plant Nutr. 56:105-109. [J,E]

Ando, J., Yuzawa, H., Mitsui, S., Kitahama, H. and Tohata, N. 1990.Composition of potassium silicate fertilizer and improvement of solubility. Jpn. J. Soil Sci. Plant Nutr. 61:282-286. [J,E]

Berthelsen, S., Noble, A. D. and Garside, A. L. 2001. Silicon research down under: Past, present, and future. pp. 241-256. In: Silicon in Agriculture, Datnoff, L. E., Snyder, G. H. and Korndorfer, G. H. eds, Elsevier Science, The Netherland.

Bowen, P. A., Menzies, J. G., Ehret, D., Samuels, L. and Glass, A. D. M. 1992. Soluble silicon sprays inhibit powdery mildew development on grape leaves. J. Amer. Soc. Hort. Sci. 17:906-912.

Calcium Silicate Fertilizer Association. 1969 Effects of calcium silicate fertilizer application on paddy rice: Experimental results conducted by national and prefectural agricultural experiment stations. 1953～1962.

Cherif, M., Asselin, A. and Belanger, R. R.1994. Defence responses induced by soluble silicon in cucumber roots infected by Pythium spp. Phytopathology 84: 236-242.

Cocker, K. M., Evans, D. E. and Hodson, M. J. 1998 The amelioration of aluminium toxicity by silicon in higher plants: solution chemistry or an in planta mechanism? Physiol. Plant 104: 608-614.

Correa-Victoria, F. J., Datnoff, L. E., Okada, K., Friesen, D. K., Sanz, J. I. and Snyder, G. H. 2001. Effecs of silicon fertilization on disease development and yields of rice in Colombia. pp. 313-323. In: Silicon in Agriculture, Datnoff, L. E., Snyder, G. H. and Korndorfer, G. H. eds, Elsevier Science, The Netherland.

Datnoff, L. E., Deren, C. W. and Snyder, G. H. 1997. Silicon fertilization for disease management of rice in Florida. Crop Prot. 16: 525-531.

Epstein, E. 1994 The anomaly of silicon in plant biology. Proc. Natl. Acad. Sci. 91: 11-17.

Epstein, E. 1999. Silicon. Annu. Rev. Plant Physiol. Plant Mol. Biol. 50: 641-664.

Frey-Wyssiling, A. 1930. Uber die Ausscheidung der Kiesel Saure in der Pflanze. Ber.Deut.Botan.Ges. 48:179-191.

Fujii, H., Hayasaka, T., Yokoyama, K. and Ando, H. 1999. Effect of silica gel application to a nursery bed of rice on rooting ability and early growth of rice plants. Jpn. J. Soil Sci. Plant Nutr. 70:785-790. [J,E]

Furue, K. and Nagata, S. 2000. Response of sugarcane to silicate soil amendment. Jpn. J. Soil Sci. Plant Nutr. 71:391-395. [J,E]

Gu, M. H., Kayama, H. and Hara, T. 1999. Effects of the supply of silicon in different forms on injury and distribution of aluminum in rice root tissues. Jpn. J. Soil Sci. Plant Nutr. 70:731-738. [J,E]

Hashimoto. H., Koda, T and Mitsui, S. 1948 Effect of application of bases and silica on the amelioration of degraded paddy soils. Jpn. J. Soil Sci. Plant Nutr. 19:61-62. [J]

Hildebrand, M., Higgins, D. R., Busser, K. and Volcani, B. E. 1993. Silicon-responsive cDNA clones isolated from the marine diatom Cylindrotheca fusiformis. Gene 132: 213-218.

Hildebrand, M., Volcani, B. E., Gassmann, W. and Schroeder, J. I. 1997. A gene family of silicon transporters. Nature 385: 688-689.

Hodson, M. J. and Evans, D. E. 1995. Aluminium/silicon interactions in higher plants. J. Exp. Bot. 46: 161-171.

Horiguchi, T. and Morita, S. 1987. Mechanism of manganese toxicity and tolerance of plants VI Effect of silicon on alleviation of manganese toxicity of barley. J. Plant Nutr. 10: 2299-2310.

Horst, W. J. and Marschner, H. 1978. Effect of silicon on manganese tolerance of bean plants (*Phaseolus vulgaris* L.). Plant Soil 50: 287-303.

Horst, W.J., Fecht, M., Naumann, A., Wissemeier, A. H. and Maier, P. (1999) Physiology of manganese toxicity and tolerance in *Vigna unguiculata* (L.) Walp. J. Plant Nutr. Soil Sci. (Z. Pflanzenernahr) 162: 263-274.

Imaizumi, K. and Yoshida, S. 1958. Edophological studies on silicon supplying power of paddy fields. Bulletin Natl. Inst. Agric. Sci. series B. No.8:261-304. [J,E]

Ishibashi, H. 1936. Influence of silica on the growth of rice plant. Jpn. J. Soil Sci. Plant Nutr. 10:244-256. [J]

Ishibashi, H. 1956. Effect of silica contained in carbonized rice husk on the growth of rice seedlings. Bulletin Fac. Agr. Yamaguchi Univ. 7: 333-340. [J, E]

Ishiguro, K. 2001. Review of research in Japan on the roles of silicon in conferring resistance against rice blast. pp. 277-291. In: Silicon in Agriculture, Datnoff, L. E., Snyder, G. H. and Korndorfer, G. H. eds, Elsevier Science, The Netherland.

Ishizuka, Y. and Tanaka, A. 1950. Effects of N, P, K on silica content of rice plant and silicified cell formation. J.Sci.Soil and Manure.Jpn. 20:138-140. [J,E]

Iwasaki, K. and Matsumura, A. 1999. Effect of silicon on alleviation of manganese toxicity in pumpkin (*Cucurbita moschata* Duch cv. Shintosa). Soil Sci. Plant Nutr. 45: 909-920.

Iwasaki, K., Maier, P., Fecht, M. and Horst, W. J. 2002a. Leaf apoplastic silicon enhances manganese tolerance of cowpea (*Vigna unguiculata*). J. Plant Physiol. 159: 167-173.

Iwasaki, K., Maier, P., Fecht, M. and Horst, W. J. 2002b. Effect of silicon supply on apoplastic manganese concentrations in leaves and their relation to manganese tolerance in cowpea (*Vigna unguiculata* (L.) Walp.). Plant Soil 238: 281-288.

Iwata, I. and Baba, I. 1962. Studies on the varietal adaptability for heavy manuring in rice. 2. Effect of silica on the adaptability of the rice plant for heavy manuring in relationship to photosynthesis. Proceedings Crop Sci. Soc. Jpn. 30:237-240. [J, E]

Izumi, K., Ishii, K. and Ando, J. 1995. Composition and solubility of potassium silicate fertilizer using Chinese coal ash. Jpn. J. Soil Sci. Plant Nutr. 66:7-12. [J,E]

Kaufman, P. B., Takeoka, Y., Carlson, T. J., Bigelow, W. C., Jones, J. D., Moore, P. H. and Ghosheh, N. S. (1979) Studies on silica deposition in sugarcane using scanning electron microscopy, energy-dispersive X-ray analysis, neutron activation analysis and light microscopy. Phytomorphology 29: 185-193.

Kato, N. 1998. Studies on the behavior of silica in paddy soil using stable isotope ^{30}Si. Radioisotopes 45:457-458. [J]

Kato, N. and Owa, N. 1996a. Dissolution of slags in water and calcium chloride solution: Effect of solution pH and calcium concentration on solubility of slags. Jpn. J. Soil Sci. Plant Nutr. 67:626-632. [J,E]

Kato, N. and Owa, N. 1996b. The factors affecting Si concentration in the soil solution: Effects of soil solution pH, Ca concentration, CO_2 gas and slag application. Jpn. J. Soil Sci. Plant Nutr. 67:655-661. [J,E]

Kato, N. and Owa, N. 1997a. Dissolution of slag fertilizers in a paddy soil and Si uptake by rice plant. Soil Sci. Plant Nutr. 43:329-341.

Kato, N. and Owa, N. 1997b. Evaluation of Si availability in slag fertilizer by an extract method using a cation exchange resin. Soil Sci. Plant Nutr. 43:351-359.

Kato, N., Yanagisawa, K. and Owa, N. 1997. Measurement of Si adsorption and 'active Si' in a soil amended with slag fertilizers by using stable isotope ^{30}Si. Soil Sci. Plant Nutr. 43: 623-631.

Kato, N. and Sumida, H. 2000. Evaluation of silicon availability in paddy soils by an extraction using a pH6.2 phosphate buffer solution. Annual Report of the Tohoku Agricultural Experiment Station for 1999 73-75. [J]

Kawamitsu, M., Kawamitsu, Y., Agata, W. and Kaufman, P. B. 1989. Effect of SiO_2 on CO_2 assimilation rate, transpiration rate, leaf conductance and dry matter production in rice plants. Sci. Bull. Fac. Kyushu Univ. 43: 161-169. [J, E]

Kawashima, R. 1927. Influence of silica on rice blast disease. Jpn. J. Soil Sci. Plant Nutr. 1:86-91. [J]

Kitada, K., Kamekawa, K. and Akiyama, Y. 1992. The dissolution of silica in soil in rotational paddy fields by the surface water dissolution method. Jpn. J. Soil Sci. Plant Nutr. 63:31-38. [J,E]

Kobayashi, J. 1954. A chemical study on the river waters of Japan. Nogaku-kenkyu 42:1-18. [J]

Kobayashi, J. 1961. A chemical study on the average quality and characteristics of river waters of Japan. Nogaku-kenkyu 48:63-106. [J]

Korndorfer, G. H. and Lepsch, I. 2001. Effect of silicon on plant growth and crop yield. pp. 133-147. In: Silicon in Agriculture, Datnoff, L. E., Snyder, G. H. and Korndorfer, G. H. eds, Elsevier Science, The Netherland.

Kumagai, K., Konno, Y., Kurada, J. and Ueno, M. 1998. Concentration of silicic acid in irrigation water on Yamagata Prefecture. Jpn. J. Soil Sci. Plant Nutr. 69:636-637. [J,E]

Liang, Y., Shen, Q., Shen, Z. and Ma, T. 1996 Effects of silicon on salinity tolerance of two barley cultivars. J. Plant Nutr. 19: 173-183.

Ma, J. F. 1988. Study on physiological role of silicon in rice plants. Master thesis, Kyoto University.

Ma, J. F. 1990. Studies on beneficial effects of silicon on rice plants. Ph. D. thesis, Kyoto University.

Ma, J. F. 2002. Beneficial Elements: Si and Na. pp. 1201-1205. In: Encyclopedia of Soil Science, Marcel Dekker, Inc. New York, USA.

Ma, J. F., Goto, S., Tamai, K., and Ichii, M. 2001. Role of root hairs and lateral roots in silicon uptake by rice. Plant Physiol. 127: 1773-1780.

Ma, J. F., Nishimura, K. and Takahashi, E. 1989. Effect of silicon on the growth of rice plant at different growth stages. Soil Sci. Plant Nutr.35: 347

Ma, J.F, Miyake, Y. and Takahashi, E. 2001. Silicon as a beneficial element for crop plants. pp. 17-39. In: Silicon in Agriculture, Datnoff, L. E., Snyder, G. H. and Korndorfer, G. H. eds, Elsevier Science, The Netherland.

Ma, J. F., Sasaki, M. and Matsumoto, H. 1997. Al-induced inhibition of root elongation in corn, *Zea mays* L. is overcome by Si addition. Plant Soil 188: 171-176.

Ma, J. F. and Takahashi, E. 1989a. Effect of silicic acid on phosphorus uptake by rice plant. Soil Sci. Plant Nutr. 35: 227-234.

Ma, J. F. and Takahashi, E.1989b. Release of silicon from rice straw under flooded conditions. Soil Sci. Plant Nutr. 35:663-667.

Ma, J. F. and Takahashi, E. 1990a. Effect of silicon on the growth and phosphorus uptake of rice. Plant Soil 126: 115-119.

Ma, J. F. and Takahashi, E. 1990b. The effect of silicic acid on rice in a P-deficient soil. Plant Soil 126: 121-125.

Ma, J. F. and Takahashi, E. 1991a. Availability of rice straw Si to rice plants. Soil Sci. Plant Nutr. 37:111-116.

Ma, J. F. and Takahashi, E. 1991b. Effect of silicate on phosphate availability for rice in a P-deficient soil. Plant Soil 133: 151-155.

Ma, J. F. and Takahashi, E. 1993. Interaction between calcium and silicon in water-cultured rice plants. Plant Soil 148:107-113.

Ma, J. F., Tamai, K., Wu, G. and Ichii, M. 2002a. A rice mutant defective in Si uptake. Submitted.

Ma, J. F. and Tamai, K. 2002b. Genotypical difference in Si uptake. Submitted.

Ma, J. F. Higashitani, A., Sato, K. and Takeda, K. 2002c. Variation in Si content of barley grain. Submitted.

Ma, J. F. and Tamai, K. 2002d. Characterization of Si uptake by rice. Submitted.

Maekawa, K., Watanabe, K., Aino, M. and Iwamoto, Y. 2001. Suppression of rice seedling blast with some silicic acid materials in nursery box. Jpn. J. Soil Sci. Plant Nutr. 72:56-62. [J,E]

Marschner, H. 1995. Mineral Nutrition of Higher Plants. 2nd edition, Academic Press p417

Matoh, T., Kairusmee, P. and Takahashi, E. 1986. Salt-induced damage to rice plants and alleviation effect of silicate. Soil Sci. Plant Nutr., 32: 295-304.

Matoh, T., Murata, S. and Takahashi, E. 1991. Effect of silicate application on photosynthesis of rice plants. Jpn. J. Soil Sci. Plant Nutr., 62: 248-251 [J, E]

Menzies, J., Bowen, P., Ehret, D. and Glass, A. 1992. Foliar applications of potassium silicate reduce severity of powdery mildew on cucumber, muskmelon and zucchini squash. J. Amer. Soc. Hort. Sci. 117: 902-905.

Meyer, J. H. and Keeping, M. G. 2001. Past, present and future research of the role of silicon for sugarcane in southern Africa. pp. 257-275. In: Silicon in Agriculture, Datnoff, L. E., Snyder, G. H. and Korndorfer, G. H. eds, Elsevier Science, The Netherland.

Mitsui, S., Hashimoto. H., and Terasawa, S. 1948. On the occurrence of rice-root rot in degraded paddy soils. Jpn.J.Soil Sci. Plant Nutr. 19:59-60. [J]

Mitsui, S., Aso, S. and Kumazawa, K. 1951. Dynamic studies on the nutrient uptake by crop plants (Part 1) Effect of H_2S on the nutrient uptake by rice roots. J. Sci. Soil and Manure. Jpn. 22:46-52.[J,E]

Mitsui, S., Kumazawa, K. and Ishibashi, T. 1953. Dynamic studies on the nutrient uptake by crop plants (Part 7) Effect of respiration inhibitors such as H_2S, NaCN, NaN_3 and butyric acid on the nutrient uptake by rice roots. J. Sci. Soil and Manure. Jpn. 24:45-50.[J,E]

Mitsui, S. and Takatoh, H. 1961. Nutritional study of silicon in graminaceous crop (Part2) Studies in the preparation of radioactive silicon-31 and its utilization for the evaluation of silicon function in crops. J. Sci. Soil and Manure. Jpn. 32:338-341.[J]

Mitsui, S. and Takatoh, H. 1962. Nutritional study of silicon in graminaceous crops (Part3). The effect of metabolic inhibitors on the silicon uptake by rice plants. J. Sci. Soil and Manure. Jpn. 33:449-452. [J]

Miyake, K. and Adachi, M. 1922. Chemische Untersuchungen uber die Widerstandigkeit der Reisarten gegen die 'Imochi-Krankheit'. J. Biochem. Japan 1:223-247. [G]

Miyake, K. and Ikeda, M. 1932. Influence of silica application on rice blast. Jpn. J. Soil Sci. Plant Nutr. 6:53-76. [J]

Miyake, Y. and Takahashi, E. 1976a. Effect silicon on the growth of silicophile plant. Comparative studies on the silica nutrition in plants (Part 9). J. Sci. Soil and Manure, Jpn. 47:375-382. [J]

Miyake, Y. and Takahashi, E. 1976b. Silicon deficiency of tomato plants (1) Demonstration of silicon deficiency. Comparative studies on the silica nutrition in plants (Part 10). J. Sci. Soil and Manure, Jpn. 47: 383-390. [J]

Miyake, Y. and Takahashi, E. 1976c. Silicon deficiency of tomato plants (2) On the environmental conditions to appearance of silicon deficiency. Comparative studies on the silica nutrition in plants (Part 11). J. Sci. Soil and Manure, Jpn. 47: 447-450. [J]

Miyake, Y. and Takahashi, E. 1976d. Silicon deficiency of tomato plants (3) On the cultivar characteristics and environmental condition. Comparative studies on the silica nutrition in plants (Part 12). J. Sci. Soil and Manure, Jpn. 47: 451-457. [J]

Miyake, Y. and Takahashi, E. 1978. Silicon deficiency of tomato plant. Soil Sci. Plant Nutr. 24: 175-189.

Miyake, Y. and Takahashi, E. 1982a. Effect of silicon on the growth of solution-cultured cucumber plant. Comparative studies on silica nutrition in plants (Part 16). Jpn. J. Soil. Sci. Plant Nutr. 53: 15-22. [J]

Miyake, Y. and Takahashi, E. 1982b. Effect of silicon on the growth of soil-cultured cucumber plant. Comparative studies on silica nutrition in plants (Part 17). Jpn. J. Soil. Sci. Plant Nutr. 53: 23-29. [J]

Miyake, Y. and Takahashi, E. 1982c. Effect of silicon on the resistance of cucumber plant to the microbial disease. Comparative studies on silica nutrition in plants (Part 18). Jpn. J. Soil. Sci. Plant Nutr. 53: 106-110. [J]

Miyake, Y. and Takahashi, E. 1983a. Effect of silicon on the growth of solution-cultured cucumber plant. Soil Sci. Plant Nutr. 29: 71-83.

Miyake, Y. and Takahashi, E. 1983b. Effect of silicon on the growth of cucumber plant in soil culture. Soil Sci. Plant Nutr. 29: 463-471.

Miyake, Y. and Takahashi, E. 1985. Effect of silicon on the growth of soybean plants in a solution culture. Soil Sci. Plant Nutr. 31: 625-636.

Miyake, Y. and Takahashi, E. 1986. Effect of silicon on the growth and fruit production of strawberry plants in a solution culture. Soil Sci. Plant Nutr. 32: 321-326.

Miyake, Y. and Takahashi, E. 1992. Mutual interaction between silicon and calcium on the growth of lowland rice in a paddy field condition. Jpn. J. Soil Sci. Plant Nutr. 63:395-402. [J,E]

Miyake, Y. and Yoneda, S. 1976a. Studies on behavior of soluble silica on polder soils in Kojima polder. Part1. Behavior of. soluble silica on paddy soils in Kojima polder. J. Sci. Soil and Manure, Jpn. 47:160-165. [J]

Miyake, Y. and Yoneda, S. 1976b. Studies on behavior of soluble silica on polder field. Part2. Behavior of silica accompanying with the cultivation. J. Sci. Soil and Manure, Jpn. 47:166-171. [J]

Miyamori, Y. 1996. Role and guideline of silicon nutrition in low protein rice production. Jpn. J. Soil Sci. Plant Nutr. 67:696-700. [J]

Mizuochi, T., Shigezumi, M., Kitsuta, Y. and Matsumoto, H.1996. Evaluation of plant available silicon in paddy soil by phosphate buffer method. Abstracts of the annual meeting, Jpn. J. Soil Sci. Plant Nutr. 42:185 [J]

Nakada, H. 1980. Mathematical statistics analysis of the influence of long term application of three nutrient elements and compost on lowland productivity. Shiga Agric. Exp. Sta. Special Bulletin. No.13:1-108. [J,E]

Ohkawa, K. 1936 Studies on plant physiological function of silica (1) Jpn. J. Soil Sci. Plant Nutr. 10:96-101. [J]

Ohta, M. 1964. Studies on slags for fertilizer use. Book 180 pages published by Kazama Shobou.[J, G]

Ohta, M. and Kobayashi, K. and Kawaguchi, U. 1953. Studies on silicates for fertilizer use. Memoirs Fac. Liberal Arts Educ. Yamanashi Univ. 4: 351-358. [J]

Ohyama, N. 1985. Effect of improvement in fertilization on the alleviation of cool-summer damage to rice. Nogyo oyobi Engei (Agriculture and Horticulture) 11:1385-1389. [J]

Okamoto, Y., Fujimaki, T. and Aoki, S. 1956. Physiological studies of the effects of silicic acid upon rice plant I. Memoirs Fac. Liberal Arts Educ. Yamanashi Univ. 7: 177-180. [J]

Okamoto, Y. 1957a. Physiological studies of the effects of silicic acid upon rice plant II. Memoirs Fac. Liberal Arts Educ. Yamanashi Univ. 8: 172-175.[J,E]

Okamoto, Y. 1957b. Physiological studies of the effects of silicic acid upon rice plant III. Proc. Crop Sci. Soc. Jpn. 25: 219-221. [J,E]

Okuda, A., Kawasaki, T. and Sakaguchi, T. 1958. Effect of calcium silicate application on the productivity of double cropping (rice-wheat) field, with special reference to nitrogen supply levels. Studies on the improvement of ultimate yields of crops by the application of silicate materials. Jpn. Assoc. for Advancement of Science 194-203. [J]

Okuda, A. and Takahashi, E. 1961a. Studies on the physiological role of silicon in crop plant. Part 1 On the method of silicon-free culture. J. Sci. Soil and Manure, Jpn. 32: 475-480. [J]

Okuda, A. and Takahashi, E. 1961b. Studies on the physiological role of silicon in crop plant. Part 2 Effect of silicon supplying period on the growth of rice plant and its nutrients uptake. J. Sci. Soil and Manure, Jpn. 32: 481-488. [J]

Okuda, A. and Takahashi, E. 1961c. Studies on the physiological role of silicon in crop plant. Part3 Effect of various amount of silicon supply on the growth of rice plant and its nutrients uptake. J. Sci. Soil and Manure, Jpn. 32: 533-537. [J]

Okuda, A. and Takahashi, E. 1961d. Studies on the physiological role of silicon in crop plant. Part 4 Effect of silicon on the growth of barley, tomato, radish, green onion, Chinese cabbage and their nutrients uptake. J. Sci. Soil and Manure, Jpn. 32: 623-626. [J]

Okuda, A. and Takahashi, E. 1962a. Studies on the physiological role of silicon in crop plant. Part5 Effect of silicon supply on the injuries of barley and rice plant due to excessive amount of Fe^{II}, Mn^{II}, Cu^{II}, As^{III}, Al^{III}, Co^{II}. J. Sci. Soil and Manure, Jpn. 33: 1-8. [J]

Okuda, A. and Takahashi, E. 1962b. Studies on the physiological role of silicon in crop plant. Part6 Effect of silicon supply on the iron uptake by rice plant from ferrous sulphate solution and the oxidation power of the root. J. Sci. Soil and Manure, Jpn. 33: 59-64. [J]

Okuda, A. and Takahashi, E. 1962c. Studies on the physiological role of silicon in crop plant. Part7 Effect of silicon supply on the growth of rice plant in various phosphorous levels. J. Sci. Soil and Manure, Jpn. 33: 65-69. [J]

Okuda, A. and Takahashi, E.1962d. Studies on the physiological role of silicon in crop plant. Part 8 Some examination on the specific behavior of low land rice in silicon uptake. J. Sci. Soil and Manure. Jpn. 33:217-221. [J]

Okuda, A. and Takahashi, E. 1962e. Studies on the physiological role of silicon in crop plant. Part9 Effect of various metabolic inhibitors on the silicon uptake by rice plant. J. Sci. Soil and Manure. Jpn. 33:453-455. [J]

Okuda, A. and Takahashi, E. 1965. The role of silicon. pp. 123-146. In: The mineral nutrition of the rice plant, The Johns Hopkins Press.

Onikura, Y. 1959. Effects of artificial silication in volcanic ash soils Part1. J. Sci. Soil Manure, Jpn. 30:357-360. [J]

Onikura, Y. 1959. Effects of artificial silication in volcanic ash soils Part 2. J. Sci. Soil Manure, Jpn. 30: 385-388. [J]

Onikura, Y. 1959. Effects of artificial silication in volcanic ash soils Part 3. J. Sci. Soil Manure, Jpn. 30: 455-458. [J]

Onodera,I. 1917. Chemical studies on rice blast (1) J. Sci. Agric. Soc. 180:606-617. [J]

Park, C. S. 2001. Past and future advances in silicon research in the Republic of Korea. pp. 359-371. In: Silicon in Agriculture, Datnoff, L. E., Snyder, G. H. and Korndorfer, G. H. eds, Elsevier Science, The Netherland.

Prabhu, A. S., Fijho, M. P. B., Filippi, M. C., Datnoff, L. E., and Snyder, G. H. 2001. Silicon from rice disease control perspective in Brazil. pp. 293-311. In: Silicon in Agriculture, Datnoff, L. E., Snyder, G. H. and Korndorfer, G. H. eds, Elsevier Science, The Netherland.

Rothamsted Experimental Station. 1978 Details of the classical and long-term experiments 1968-73.

Sachs, J. 1862. Ergebnisse einiger neuerer untersuchungen uber die in Pflanzen enthaltene Kieselsaure. Flora S53 cited from Wagner, F.1940.

Saigusa, M.,Yamamoto, A. and Shibuya, K. 1998a. Agricultural use of porous hydrate calcium silicate (1) Evaluation of porous hydrate calcium silicate as a silicate fertilizer. Jpn. J. Soil Sci. Plant Nutr. 69:576-581. [J,E]

Saigusa, M., Yamamoto, A. and Shibuya, K. 1998b. Agricultural use of porous hydrat calcium silicate (2) Practical use to improve silicon nutrition of rice plants in paddy fields. Jpn. J. Soil Sci. Plant Nutr. 69:612-617. [J,E]

Sangster, A. G. 1978. Silicon in the roots of higher plants. Am. J. Bot. 65:929-935.

Sasamoto, K. 1958. Studies on the relation between the silica content in the rice plant and insect pests VI. On the injury of silicated rice plant caused by the rice stem borer and its feeding behavior. Jpn. J. Appl. Entomology and Zoology. 2:82-92. [J.E]

Sasamoto, K 1960. Studies on the relation between the silica content in the rice plant and insect pests VIII. Feeding preference of the rice stem borer larvae for the rice plants cultured in soil of different nitrogen levels. Jpn. J. Appl. Entomology and Zoology. 4:115-118. [J,E]

Sasamoto, K 1961. Resistance of the rice plant applied with silicate and nitrogenous fertilizers to the stem borer *Chilo suppressalic* Walker. Proceedings of the Faculty of Liberal Arts and Education. Yamanashi Univ. No3 1-73. [J,E]

Savant, N. K., Snyder, G. H. and Datnoff, L. E. (1997) Silicon management and sustainable rice production. Advan. Agron. 58: 151-199.

Seo, S. W. and Ohta, Y. 1982. Role of the hull in the ripening of rice plant V. Water loss in hull and development of rice kernel. Jpn. J. Crop Sci. 51: 529-534.[J]

Shimoyama, S. 1958. Effect of calcium silicate application to rice plant on the alleviation of lodging and damage from strong gale. Studies on the improvement of ultimate yields of crops by the application of silicate materials. Jpn. Assoc. for Advancement of Science, 57-99.[J]

Strigel, 1912. Mineralstoff aufnahme verschiedener Pflanzenarten aus ungedungtem Boden. Landwirtsch Jb. 43: 349-371.

Sujatha, G., Reddy, G. P. V. and Murthy, M. M. K. 1987. Effect of certain biochemical factors on expression of resistance of rice varieties to brown planthopper (*Nilaparvata lugens* Stal). J. Res. Andhra. Pradesh Agric. Univ. 15: 124-128.

Sumida, H. 1991. Characteristics of silica dissolution and adsorption in paddy soils. Application to soil test for available silica. Jpn. J. Soil Sci. Plant Nutr. 62:378-385. [J,E]

Sumida, H. 1996. Progress and prospect of the research on paddy soil management under various rice growing system. 1 Progress in nutrient behavior and management research on paddy soil (3) silicic acid. Jpn. J. Soil Sci. Plant Nutr. 67:435-439. [J]

Sumida, H. and Ohyama, N. 1988. Silica content in the supernatant obtained by the incubation under submerged conditions. Jpn. J. Soil Sci. Plant Nutr. 59:593-599. [J,E]

Sumida, H. and Ohyama, N. 1991. The effects of application of organic matters and calcium silicate on silica uptake by rice plant. Jpn. J. Soil Sci. Plant Nutr. 62:386-392. [J,E]

Snyder, G. H. 1991. Development of a silicon soil test for Histosol-grown rice. Belle Glade EREC Research Report EV-1991-2, pp. 29-39. University of Florida, Belle Glade, USA.

Snyder, G. H., Jones, D. . and Gascho, G. J. 1986. Silicon fertilization of rice on everglades Histosols. Soil Sci. Soc. Am. J. 50: 1259-1263.

Takahashi, E. 1966a. Studies on the physiological role of silicon in crop plant. Part13 Effect of silicon on the resistance to radiation injury. J. Sci. Soil and Manure, Jpn. 37: 183-188.[J]

Takahashi, E. 1974. Effect of soil moisture on the uptake of silica by rice plant seedlings. J. Sci. Soil and Manure, Jpn. 45:591-596.[J]

Takahashi, E. 1982. Effect of co-existing inorganic ions on silicon uptake by rice seedling. Comparative studies on silica nutrition in plants (Part20). Jpn. J. Soil Sci. Plant Nutr. 53:271-276. [J]

Takahashi, E., Arai, K. and Kashida, Y. 1966b. Studies on the physiological role of silicon in crop plant. Part 14 Effect of silicon on the assimilation and translocation of $^{14}CO_2$ supplied at the various growth stages of rice. J. Sci. Soil and Manure, Jpn. 37: 594-598. [J]

Takahashi, E. and Hino, K. 1978. Silica uptake by plant with special reference to the forms of dissolved silica. J. Sci. Soil and Manure. Jpn. 49:357-360. [J]

Takahashi, E., Ma, J. F. and Wang, D. P. 1990. Research on silicon fertilization in China. Agriculture and Horticulture, 65: 25-30. [J].

Takahashi, E., Matsumoto, H., Syo, S. and Miyake, Y. 1976. Effect of germanium on the growth of plants with special reference to the silicon nutrition (Part 3). J. Soil Sci. Manure, Jpn. 47: 217-221. [J]

Takahashi, E. and Miyake, Y. 1976a. Distribution of silica accumulator plants in the plant kingdom (1) Monocotyledons. J. Sci. Soil and Manure. Jpn. 47:296-300.[J]

Takahashi, E. and Miyake, Y. 1976b. Distribution of silica accumulator plants in the plant kingdom (2) Dicotyledons. J. Sci. Soil and Manure. Jpn. 47:301-306. [J]

Takahashi, E.and Miyake, Y. 1976c. Distribution of silica accumulator plants in the plant kingdom (3) Gymnospermae, Pteridophyta and Bryophyta. J. Sci. Soil and Manure. Jpn. 47:333-337.[J]

Takahashi, E. and Nishi, T. 1982 Effects of nutritional conditions on the silicon uptake by rice plants. Comparative studies on silica nutrition in plants (Part21). Jpn. J. Soil Sci. Plant Nutr. 53:395-401. [J]

Takahashi, E., and Okuda, A. 1963a. Studies on the physiological role of silicon in crop plant. Part 10 Effect of some sugars and organic acids on silicon uptake by rice plant. J. Sci. Soil and Manure, Jpn. 34: 114-118. [J]

Takahashi, E., and Okuda, A. 1963b. Studies on the physiological role of silicon in crop plant. Part 11 Effect of light on the silicon uptake by the seedlings of rice plant. J. Sci. Soil and Manure, Jpn. 34: 397-402. [J]

Takahashi, E., Syo, S. and Miyake, Y. 1976a. Effect of germanium on the growth of plants with special reference to the silicon nutrition (Part 1). J. Soil Sci. Manure, Jpn. 47: 183-190. [J]

Takahashi, E., Syo, S. and Miyake, Y. 1976b. Effect of germanium on the growth of plants with special reference to the silicon nutrition (Part 2). J. Soil Sci. Manure, Jpn. 47: 191-197. [J]

Takahashi, E., Tanaka, T. and Miyake, Y. 1981a. Distribution of silica accumulator plants in the plant kingdom (4) Pteridophyta. Jpn. J. Soil Sci. Plant Nutr. 52:445-449. [J]

Takahashi, E.,Tanaka, T. and Miyake, Y. 1981b. Distribution of silica accumulator plants in the plant kingdom (5) Gramineae. Jpn. J. Soil Sci. Plant Nutr. 52: 503-510. [J]

Takahashi, E., Tanaka, T. and Miyake, Y. 1981c. Distribution of silica accumulator plants in the plant kingdom (6) Commelinales, Cyperales and some Dicotyledoneae families (Cucurbitaceae and Urticaceae). Jpn. J. Soil Sci. Plant Nutr. 52:511-515. [J]

Takahashi, K. 1981. Effects of slags on the growth and silicon uptake by rice plants and the available silicates in paddy soils. Bulletin Shikoku Agric. Exp. Sta. 38:75-114. [J,E]

Takahashi, K. and Nonaka, K. 1986. Method of measurement of available silicate in paddy field. Jpn. J. Soil Sci. Plant Nutr. 57:515-517. [J]

Takeoka, Y., Matsumura, O. and Kaufman, P. B. 1983. Studies on silicification of epidermal tissues of grasses as investigated by soft X-ray image analysis I. On the method to detect and calculate frequency of silica bodies in bulliformcells. Jpn. J. Crop Sci. 52: 544-550.[J]

Tsuda, M., Morita, M., Makihara, D. and Hirai, Y. 2000. The involvement of silicon deposition in salinity-induced white head in rice (*Oryza sativa* L.). Plant Production Sci. 3: 328-334.

Tsuno, Y. and Higashi, K. 1984a. Studies on the antagonism between Si and Ca in rice plants.I. Effect of wind speed and water depth of irrigation on Si and Ca content of leaf blade. Jpn. J. Crop Sci. 53:76-77. [J]

Tsuno, Y. and Higashi, K. 1984b. Studies on the antagonism between Si and Ca in rice plants. II Antagonism between the content of Ca and Si in the leaf blades. Jpn. J. Crop Sci. 53:78-79. [J]

Tsuno, Y. and Kasahara, H. 1984. Studies on the antagonism between Si and Ca in rice plants. III Effects of Si and Ca on the resistance of rice plants to blast disease in the field. Jpn. J. Crop Sci. 53:76-77. [J]

Ueda, K. and Ueda, S. 1961. Effect of silicic acid on bamboo growth. Bulletin Kyoto Univ. Forests. 33: 79-99 [J, E]

Voogt, W. and Sonneveld, C. 2001. Silicon in horticultural crops grown in soilless culture. pp. 115-131. In: Silicon in Agriculture, Datnoff, L. E., Snyder, G. H. and Korndorfer, G. H. eds, Elsevier Science, The Netherland.

Wagner, F. 1940. Die Bedeutung der Kieselsaure fur das Wachstum einiger Kulturpflanzen, ihren Nahrstoff-haushalt und ihre Anfalligkeit gegen echte Mehltaupilze. Phytopath. Z. 12: 427-479.

Wang, H., Li, C. and Liang, Y. 2001. Agricultural utilization of silicon in China. pp. 343-358. In: Silicon in Agriculture, Datnoff, L. E., Snyder, G. H. and Korndorfer, G. H. eds, Elsevier Science, The Netherland.

Williams, D. E. and Vlamis, J. 1957. The effect of silicon on yield and Mn-54 uptake and distribution in the leaves of barley plants grown in culture solutions. Plant Physiol. 32: 404-409.

Woolley, J. T. 1957. Sodium and silicon as nutrients for the tomato plant. Plant Physiol. 32: 317-321.

Yamanaka, R. and Sakata, M. 1993. Singularity of silicic acid absorption and manganese toxicity on cucumber grafted bloomless stock. Jpn. J. Soil Sci. Plant Nutr. 64:319-324. [J]

Yamanaka, R. and Sakata, M. 1994. The countermeasures and characteristics of manganese excess toxicity occurred in cucumbers grafted on bloomless stocks. Jpn. J.Soil Sci. Plant Nutr. 65:337-340. [J]

Yeo, A. R., Flowers, S. A., Rao, G., Welfare, K., Senanayake, N. and Flowers, T. J. 1999 Silicon reduces sodium uptake in rice (*Oryza sativa* L.) in saline conditions and this is accounted for by a reduction in the transpirational bypass flow. Plant Cell Environ. 22: 559-565.

Yoshida, S. 1965. Chemical aspects of the role of silicon in physiology of the rice plant. Bull. Natl. Inst. Agric. Sci. Series B. 15: 1-58. [J. E.]

Yoshida, S., Navasero, S. A. and Ramirez, E. A. 1969. Effects of silica and nitrogen supply on some leaf characters of the rice plant. Plant Soil 31: 48-56.

Yoshida, S., Ohnishi, Y. and Kitagishi, K. 1959a. The chemical nature of silicon in rice plant. Soil and Plant Food 5: 23-27.

Yoshida, S., Ohnishi, Y. and Kitagishi, K. 1959b. Role of silicon in rice nutrition. Soil and Plant Food 5: 127-133.

Yoshida, S., Ohnishi, Y. and Kitagishi, K. 1962a. Histochemistry of silicon in rice plant I. A new method of determining the localization of silicon within plant tissues. Soil Sci. Plant Nutr. 8: 30-35.

Yoshida, S., Ohnishi, Y. and Kitagishi, K. 1962b. Histochemistry of silicon in rice plant II. Localization of silicon within plant tissues. Soil Sci. Plant Nutr. 8: 36-41.

Yoshida, S., Ohnishi, Y. and Kitagishi, K. 1962c. Histochemistry of silicon in rice plant III. The presence of cuticule-silica double layer in the epidermal tissue. Soil Sci. Plant Nutr. 8: 1-5.

Yoshida, S., Ohnishi, Y. and Kitagishi, K. 1962d. Chemical forms, mobility and deposition of silicon in rice pant. Soil Sci. Plant Nutr. 8: 15-21.

Yoshii, H., Nonaka, H. and Iwata, T. 1958. The effect of silica supply on the severity of rice stem rot caused by *Leptosphaeria salvinii* Cat. Studies on the improvement of ultimate yield of crops by the application of silicate materials. Jpn. Assoc. for Advancement Science. 37-41.[J, E]

*[J], [J,E] at the end of the reference indicates papers written in Japanese, in Japanese with English summary, respectively.

Index

N

O

P

T

U

V

W

Y

Z

www.ingramcontent.com/pod-product-compliance
Ingram Content Group UK Ltd.
Pitfield, Milton Keynes, MK11 3LW, UK
UKHW020245290926
15470UKWH00002B/761

9 780444 511669